THE NEUROSCIENCE OF SPIRITUAL TRANSFORMATION

Ephraim T. Gwebu

Copyright © 2022 *Ephraim T. Gwebu*

All rights reserved.

Contents

Dedication i

Acknowledgments ii

Foreword iii

Acclaim v

Introduction vi

Abbreviations Of Bible References: viii

Chapter 1 The War-Within For A Typical Christian 1

Chapter 2 The Human Brain, The Pinnacle Of God's Creation 6

Chapter 3 Potter's Clay And Neuroplasticity 11

Chapter 4 The Neuroscience Of A Christ Orchestrated Heart Transplant 15

Chapter 5 Neuroscience Of New Creation In Christ 18

Chapter 6 The Neuroscience Of Mindset 22

Chapter 7 The Neuroscience Of Mindfulness 27

Chapter 8 The Neuroscience Of Mindsight 32

Chapter 9 The Neuroscience Of Visualization (Mental Rehearsal) . 36

Chapter 10 Neuroplasticity And Self-Talk 42

Chapter 11 Humility: A Source Of Christian Joy And The Opposite Of Pride 50

Chapter 12 The Neuroscience Of Kindness 61

Chapter 13 The Neuroscience Of Habits 66

Chapter 14 Wandering And Distractibility: Enemies Of Perfect Peace In Jesus 71

Chapter 15 Guilt: A Cause Of Chronic Stress In A Christian 76

Chapter 16 Iniquity In Leadership 83

Chapter 17 The Neuroscience Of Leviticus 19:17, 18 86

References 96

Dedication

To the LORD Jesus Christ, Creator of the human brain, the most complex system on earth

To my dear and loving wife, Keratiloe

To our beloved daughter and son Nomalanga Thenjiwe and Thabani Musa

And in Loving Memory of our Parents who sacrificed so much for us:

Tobela Mayenga Gwebu   *Swelibizo Ndlalambi Gwebu*

Andries A. Noko   *Abigail Noko*

Acknowledgments

I am indebted to Dad, who used the Bible to teach me at age 4, the vowels of the alphabet that gave rise to my making it my most favorite book. My mom instilled the desire to learn even though she did not go beyond 5th grade. My parents-in-law demonstrated their trust and dependence on God as they always said: 'Hare Rapeleng' (Let us pray) out of habit. I would like to thank my dear wife, my encourager and objective critic, for her intuitive review of the manuscript. My daughter, Nomalanga, asked intellectually challenging question during the early stages of the writing. My son, Thabani, designed the book cover.

I acknowledge, with appreciation, the following friends (i) Gregory and Carol Allen, for making sure that Jesus Christ is central in the book, (ii) Craig and Janis Newborn, for insisting that the Holy Spirit must be seen to play a major catalytic role in spiritual transformation. My sincere gratefulness is extended to Dr. Ishmael Noko for the acclaim and Dr. Clifford Sibanda for the foreword. Finally, I am grateful to the LORD for permitting us to serve at Solusi and Rusangu Universities, which motivated us to write this book.

Foreword

Spiritual formation is usually presented as a mystical process that has no connection to our understanding of the human anatomy and physiology. This book provides groundbreaking information that explains spiritual formation from a neurological standpoint. Ephraim Gwebu demonstrates the meaning of behaviour change from key biblical passages. While Gwebu exhibits a deep understanding of these neurological processes, he writes clearly and simply so that even those without a scientific background can follow the line of argument. This is the Bible made exciting and easy. This book is a must read. It dovetails with different branches of theology. For those in Biblical theology, this book provides new avenues of reflecting on biblical passages. This is a book that provokes a Systematic theologian to reflect on topics like repentance, justification, sanctification, and sin. For those in Applied Theology, this book presents insights that can revolutionize our homiletics, counselling, church administration, and mission. It is not just another book on the shelf, this is the book that has been missing. The utility of this book is not limited to the seminary. This is the right tool for any parent who needs to understand the dynamics of behaviour formation among children. For those planning to conduct seminars on behaviour, this is a valuable resource. The book does not merely end with a list of answers, it provokes the right questions. This is a book that will provoke a meaningful conversation on behaviour change.

Clifford Sibanda, Ph.D., Professor, Faculty of Theology and Chaplaincy, Solusi University, P.O. Box Solusi, Bulawayo, Zimbabwe.

Acclaim

In the multidisciplinary field of neurotheology, known as spiritual neuroscience, is the fundamental attempt to understand the relationship between the human brain and religion. In this book, Professor Ephraim Gwebu makes a punctuating contribution. He demonstrates how neuroscience and neuroscientists can be resourceful for biblical exegesis. The book can also be a resource for pastors, theologians, early childhood development practitioners and many others who are carrying the responsibility of transforming humans and human behavior. I am not only excited about the publication of this book, but I am thankful to the author for enriching my own understanding of how neuroscience can be so resourceful to Christian religion.

Ishmael A. Noko, Ph.D., Founder and CEO of Lutheran University of Southern Africa and Former Secretary-General (President) of the Lutheran World Federation.

Introduction

The Neuroscience of Spiritual Transformation is an attempt to reveal the field of neuroscience as evidence of God's handiwork in creating the brain to be the most complex system on earth. The book shows how the brain may be a conduit for repentance, sanctification, and spiritual transformation. I believe God created human beings in anticipation that they would sin against Him. Therefore, He made a human brain that can be molded like a potter's clay. The brain is malleable. This book shows how this property of the brain may be instrumental in spiritual transformation, including hating the only thing that the LORD hates, namely, sin. The average Christian lives a life of contradictions in that they know the right thing to do, but they do not do it.

The book shows that when the Christian focus is on Jesus Christ alone, spiritual transformation is inevitable. The book illustrates how the average Christian can focus on Jesus exclusively, using the brain's malleability quality. Also described in the book is that when our minds are focused on Christ with intentions to please Him, the Holy Spirit changes/molds the brain's networks, like a potter working with clay, to produce God-pleasing Christians

Christians are called upon to imitate Christ Jesus and visualize His glory and goodness. The book describes the neuroscience of visualization and envisioning. The process stimulates the brain to produce nerve cells called mirror neurons. I propose that visualizing and envisioning the LORD produces Christ-centered mirror neurons so that Christ's presence is felt at all times in the mind of the Christian.

The author is a Christian of 48 years. He holds a Ph.D. in Biochemistry from The Ohio State University, and has served as both university professor and administrator. He is a member of the Society for Neuroscience and the American Society for Biochemistry and Molecular Biology.

Abbreviations Of Bible References:

AMP: Amplified Bible

BSB: Berean Study Bible

CJB: Complete Jewish Bible

CSB: Christian Standard Bible

ERV: Easy-to-Read Version

ESV: English Standard Version

HCSB: Holman Christian Standard Bible

KJV: King James Version

MEV: Modern English Version

NASB: New American Standard Bible

NET: New English Translation

NIV: New International Version

NKJV: New King James Version

NLT: New Living Translation

Chapter 1
The War-Within For A Typical Christian

Many of us Christians suffer from the guilt of knowing that we are not following all that the LORD Jesus Christ has commanded. He says, "Do unto others as you would like them to do unto you." (Matthew 7:12, KJV) Do we, as Christians, treat others the way that we would like to be treated? Also, the LORD has commanded us: "Do not seek revenge or bear a grudge against anyone among your people, but love your neighbor as yourself. I am the Lord." (Leviticus 19:18, NIV) To what extent do Christians worldwide follow this command? We know the command but do not follow it. Thus, we are in a continuous struggle with ourselves, which results in the absence of love, joy, and peace.

The apostle Paul describes himself as "an apostle of Jesus Christ by the will of God." (2 Corinthians 1:1, NIV). In Romans 1:1, he identifies himself as "Paul, a servant of Jesus Christ, called to be an apostle unto the gospel of God." Yet he declares, "I don't really understand myself, for I want to do what is right, but I don't do it. Instead, I do what I hate." (Romans 7:15, NLT). Thus, he identifies the conundrum that many Christians face.

Since the LORD created us, there has to be some way that we can follow Him as He commands. Jesus asks us to abide in Him. How can we do this? When He created us, could He have installed a system that enables us to do as He commands? I believe that the LORD did exactly that.

This book demonstrates how discoveries in neuroscience (study of the brain) reveal one of these systems that can help us in our struggle with Biblical texts such as the following:

"Turn! Turn from your evil ways! Why will you die..?" (Ezekiel 33:11, NIV)

"I will take away your stubborn heart and give you a new heart and a desire to be faithful. You will have only pure thoughts." (Ezekiel 36:26, CEV).

A New Creation In Christ

"Therefore if any man is in Christ, he is a new creature: old things are passed away; behold, all things have become new." (2 Corinthians 5:17, NIV).

The Mind of Christ

"Let this mind be in you which was also in Christ Jesus." (Philippians 2:5, NKJV).

"Do not be conformed to this present world, but be transformed by the renewing of your mind, so that you may test and approve what is the will of God—what is good and well-pleasing and perfect." (Romans 12:2, NET)

Be Perfect as the Father in Heaven is Perfect

"You, therefore, will be perfect [growing in spiritual maturity both in mind and character, actively integrating godly values into your daily life], as your heavenly Father is perfect."

(Matthew 5:48, Amplified Bible, AMP); "You, therefore, must be perfect [growing in complete maturity of godliness in mind and character, having reached the proper height of virtue and integrity], as your heavenly Father is perfect." (Matthew 5:48 Amplified Bible, Classic Edition - AMPC)

Have Perfect Peace

"You will keep him in Perfect Peace whose mind is stayed on You because he trusts in you." (Isa. 26:3, ESV)

Be Filled with Praise!

"Praise the LORD, O my soul; all my inmost being, praise His holy Name." (Psalm 103:1, NIV)

Be a Consistent Christian

"Therefore, my dear friends, as you have always obeyed—not only in my presence, but now much more in my absence—continue to work out your salvation with fear and trembling..." (Phil. 2:12, NIV)

"For as he thinketh in his heart, so is he." (Proverbs 23:7, KJV)

Jesus said, "...Follow Me!" (John 21:19 NIV). Paul said, "Imitate me, as I also imitate Christ." (I Corinthians 11:1, Christian Standard Bible)

The intent of the book is to show that when God created us, He included built-in mechanisms to make it possible for sinful human beings to completely surrender to Him. For example, He tells us to be kind to others. When we show kindness, the brain produces a kindness hormone that makes the kind person feel warm and cuddly. The more kindness we show, the better we feel. Thus, the hormone facilitates repetition of the act of kindness. Scientists have discovered that visualization or having a mental rehearsal, produces networks for this process. These networks can be strengthened like muscles. I believe that when we visualize the goodness of the LORD, the brain networks attuned to serving God are formed. These God-serving networks are strengthened by repetition. The Holy Spirit facilitates this process.

The Role of the Holy Spirit in Spiritual Transformation

All Biological processes occur as reactions converting A to B. For example, when sugar (glucose) is being broken down to produce energy, there is a series of reactions involved. Each one of the reactions cannot occur without the involvement of a catalyst (facilitator). It is proposed that spiritual transformation is a biological reaction. When one surrenders to God, the Holy Spirit is the catalyst in the formation of God-serving brain networks, where 'A' is the old nature and 'B' in the new creation.

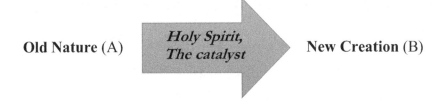

Old Nature (A)   *Holy Spirit, The catalyst*   **New Creation (B)**

In chemistry, there are 2 types of reactions. There are reactions that occur spontaneously. Then there are those that require catalysts for them to occur. Most biological reactions are not nonspontaneous and require biological catalysts, called enzymes, to occur. I propose that spiritual transformation is a biological reaction and occurs under the catalysis (enzyme action) of the Holy Spirit. Human beings are incapable of facilitating spiritual transformation on their own strength. The Holy Spirit is the enzyme. All that a human being has to do is to surrender and heed the voice of the LORD. The Holy Spirit is the transforming agent acting through neuroscientific processes described in this book. Jesus said: "…. apart from Me you can do nothing" (John 15:5, NIV). Jesus further said: "Behold, I stand at the door and knock. If anyone hears My voice and opens the door, I will come in and dine with him, and he with Me" (Revelation 3:20, BSV). When we open the door (brain/mind), the Holy Spirit catalyzes the spiritual transformation (creation of Christ-centered brain networks).

This book attempts to show that the human brain created by God has built-in mechanisms that are channels for Christians to please the LORD and hate what He hates under the catalysis of the Holy Spirit. Thus, Spiritual transformation is a process for structuring and and restructuring these mechanisms that make Jesus Christ the center of the brain universe of a Christian based on the brain's amazing capacity to change and adapt to new information and experiences.

Chapter 2
The Human Brain,
The Pinnacle Of God's Creation

Genesis 1:26 says, "Then God said, 'Let us make human beings in our image, to be like us…'" (New Living Translation, NLT). One of the greatest wonders and mysteries of the human body is the brain. There is nothing on earth that is more complex. This three-pound (1.34 kg) organ is the center of intelligence and controls everypart of the body. It interprets all the senses, initiates body movement, and controls behavior. God encapsulated the brain in a bony shell to protect and bathe it in an internal protective fluid and an array of intricate blood vessels. It produces thoughts, makes decisions, sets goals, forms memories, gives us our personalities, regulates our emotions, controls movement, and allows us to interact with the world around us. The brain was designed to be the source of all the qualities that define our humanity, making the brain the pinnacle of God's earthly creation. Its complexity continues to baffle neuroscientists worldwide.

Neuroscience is the study of how the nervous system (brain and spinal cord) develops, its structure, and how it functions. Neuroscientists focus on the brain and its impact on behavior and cognitive functions. The nervous system is the center of all mental activity, including thought, learning, and memory. Thus, it is the major controlling, regulatory, and communicating system in the body. The mind is described as the part of a person that thinks, reasons, feels, understands, remembers, and responds as part of brain function.

Neurons (also called nerve cells) are the basic units of the brain. Neurons receive sensory input from the external and internal world. For example, they send motor commands to our muscles to act, interpret what our ears hear and what our eyes see, and coordinate all the necessary functions of life. Their interactions define who we are as people created in the image of God. Neurons make this possible through a highly complex and sophisticated network of fibers called neural pathways. They carry information throughout the body using electrical and chemical signals (messages), coordinating all the functions of life. They detect what is going on around and within us. They decide how we should act, change the state of internal organs (heart rate, digestion, breathing, etc.), and allow individuals to think and remember what is going on around them.

The human brain is the most complex structure in the universe (https://mindmatters.ai/2022/03/yes-the-human-brain-is-the-most-complex-thing-in-the-universe/). It is a universe within itself. The brain universe is called Neuroverse™ (Douyon, 2019), consisting of over one hundred billion neurons with supporting cells called glial cells. Philippe Douyon, a neurologist, says that our Neuroverse™ has the ability to create new neurons, build new connections, strengthen old ones, and prune neurons that are no longer in use. (Douyon, 2019).

The Neuroverse™ is malleable. Everything that we experience changes the brain. The Neuroverse™ is influenced by every thought we have, every challenge we encounter, every new perspective we become aware of, as well as every habit we develop. The Neuroverse™ is capable of undergoing remarkable changes called neuroplasticity. Whenever we decide to learn something, we create new neurons that form neural networks for that learning. This process of neuroplasticity occurs throughout the life-span. Neural networks are connected neurons. Each neuron is connected to as many as 1,000 other neurons, creating an incredibly complex network of communication. This is a manifestation of God's Majestic Handiwork.

The term "neuroplasticity" describes how experiences reorganize (rewire) neural pathways in the brain throughout one's lifetime. When we learn new things, memorize new information, or experience new occurrences, long-lasting functional changes occur in the brain's neural connections because of neuroplasticity.

To illustrate plasticity, imagine making an impression of your thumb in a lump of malleable clay. For the thumb's impression to appear in the clay, changes must occur in the clay itself. The shape of the clay changes as the thumb presses into it. Correspondingly, the neural network in the brain must reorganize (rewire) in response to a new experience. Since the brain is malleable, life experiences are 'imprinted' in it, so to speak. It is this process of continuous adaptation that allows us to learn throughout our lives. Thus, each experience or learning process involves the formation of a specific neural pathway.

Neuroscientists tell us that for every thought we have, every action that we take, our dreams, our emotions, and our connections to the people that we love or care about is a result of neural connections (pathways) that are created for each activity, regardless of age. Moses admonished parents to teach their children God's Commandments: "Repeat them again and again to your children. Talk about them when you are at home and when you are on the road, when you are going to bed and when you are getting up." (Deut. 6:6,7, NLT). Moses was encouraging parents to help young children develop neural pathways in their brains for serving the LORD.

Adult brains are also capable of neuroplastic adaptation. Research shows that the brain never stops changing in response to learning. Our brains continue to create new neural pathways and alter or eliminate existing ones to adapt to new experiences, as well as acquire and learn new information. There is a strong possibility that constant exposure to Godliness, in cooperation with the Holy Spirit, can create permanent neural pathways that displace those for selfishness, pride, hatred, anger, malice, and filthy communication. The new neural pathways give the learner and practitioner "bowels of mercy, kindness, humbleness of mind, meekness, longsuffering, forbearing one another, and forgiving one another." (Colossians 3:12, KJV). I believe that our sin-prone brains can be rewired through neuroplasticity to be Christians that please the LORD Jesus through practices that stimulate the formation of new Christ-centered neural pathways.

In this book, an effort is made to illustrate how we can (i) be reshaped like the potter's clay in the hands of Jesus, (ii) abide in Christ, (iii) have the mind of Christ, (iv) be transformed by the renewing of our minds, (v) be perfect as our Father is perfect, and (iv) have perfect peace through the amazing process of neuroplasticity that was implanted by the LORD Jesus in Genesis 1:26, 27. Based on discoveries in neuroscience, we now have a better idea of how the LORD Jesus Christ designed our brains to make it possible for us to choose to follow Him and for the Holy Spirit to superintend the structural and functional changes in the brain, supplanting old evil tendencies with desires to serve Him.

Thus, this book uses the principles of neuroscience to show how we can follow God's commands and make pleasing the LORD a habit. The book also endeavors to show how this pinnacle of God's creation, the human brain, may be a conduit through which the Holy Spirit works to attenuate and mitigate the Christian struggles of knowing what is right but failing to do so.

Chapter 3
Potter's Clay And Neuroplasticity

"Turn! Turn from your evil ways! Why will you die...?" (Ezekiel 33:11 NIV)

Growing up in a village, pottery craft was a major activity for making cooking pots of various sizes. I was often intrigued by the potter who would redo the marred pot several times until the final product was just right. The clay had to be the right texture and malleability.

Figure 1. Clay in the Potter's Hand

Source: https://www.cslewisinstitute.org/resources/pride-and-humility

Because of this experience in my village and because of my familiarity with the potter's work, Jeremiah chapter 18 has often intrigued me. "This is the word that came to Jeremiah from the Lord: 'Go down to the potter's house, and there I will give you My message.' So I *(Jeremiah)* went down to the potter's house, and I saw him working at the wheel. But the pot he was shaping from the clay was marred in his hands, so the potter formed it into another pot, shaping it as seemed best to him. Then the word of the LORD came to me. He said, "Can I not do with you, Israel, as this potter does?' declares the Lord. 'Like clay in the hand of the potter, so are you in My hand,

Israel." (Jeremiah 18:1-6, NIV). God created our brain to be malleable in His hand as the clay in the potter's hands.

I have often wondered about the meaning and practicality of us being clay in the LORD's hands, that He can shape into what seemed best to Him. The LORD is saying that He can mold (shape) each one of us to what seems best to Him. A recent discovery in the study of the brain (neuroscience) is that each of us has a brain that is malleable/plastic and can be molded by what we do, experience, learn, practice, and think. I propose that spiritual transformation is allowing the LORD Jesus to mold us into His likeness through the neuroplastic brain that He created since our brains are capable of creating new neural pathways and altering or eliminating existing ones. That may be why the LORD declared in Isa 1:18. Come now, and let us reason together, saith the LORD: though your sins are as scarlet, they shall be as white as snow; though they are red like crimson, they shall be as wool" (KJV). The LORD is saying that He can replace the crimson (sinful) neural pathways with new Christ-like neural pathways. I believe that this is possible because of neuroplasticity, a mechanism created by God to make this possible.

In Genesis 3:8, the Bible says that the LORD God walked in the garden of Eden in the cool of the day, looking for Adam and Eve. I believe that this encounter with the LORD occurred regularly, creating neural pathways in Adam and Eve of His goodness and love. However, on this day, Adam and Eve had listened to Satan's lies, which created neural pathways of doubt, dishonesty, disobedience, and fear. The result was an effort by Adam and Eve to hide from God.

I believe that intimate, on-going exposure to Godliness creates permanent neural pathways that displace those described by Paul in Colossians 3:8: "But now you must also rid yourselves of all such things as these: anger, rage, malice, slander, and filthy language from your lips." (NIV). New neural pathways, implanted by the Holy Spirit, give the penitent learner/practitioner "bowels of mercies, kindness, humbleness of mind, meekness, longsuffering; forbearing one another, and forgiving one another..." (Colossians 3:12,13 KJV).

After his encounter with the LORD, Isaiah exclaimed: "Woe is me! For I am lost; for I am a man of unclean lips, and I dwell among a people of unclean lips; for my eyes have seen the King, the LORD of hosts!" (Isa 6:5, RSV). An angel touched Isaiah's lips with hot coal and assured him: "...your guilt is taken away, and your sin forgiven." (Isa 6:7, RSV). Likewise, with us, an intimate encounter with the LORD facilitates the rewiring of our plastic brains so that we recognize our guilt, seek forgiveness, and become submissive to His will for us. Isaiah's encounter with the LORD produced neural pathways that made him recognize his unworthiness and exclaim, "I am lost, for I am a man of unclean lips." This was a spiritual enlightenment. Figure 2 is a proposal for a path to a new creation in Potter's hand, a spiritually transformed Christian.

Figure 2. Clay in the Potter's (Jesus Christ) Hands:

God's call in Eze. 33:11 to turn from our wicked ways is a request for us to direct and focus our thoughts and intentions toward pleasing the LORD. Submitting to the authority of the Potter allows Him to create God-pleasing neural pathways in the brain, supplanting the old sinful ones. Jesus is the Potter. The clay is our neuroplastic brain. Our part is to submit our hearts and minds to the Potter to be molded by Him, and intentionally focus our thoughts on His righteousness, His self-sacrifice, and His gift of salvation. Jesus Christ, the Potter, produces new God-fearing neural pathways that replace the old ones. Our response is to love the LORD Jesus and accept His transformation of our minds. Like clay in the hands of the Potter, the brain, under the influence of the Holy Spirit, responds to its Creator by producing neural networks that become the channel for a Christ-like life.

Chapter 4
The Neuroscience Of A Christ Orchestrated Heart Transplant

In Ezekiel 36:26, the LORD says, "I will take away your stubborn heart and give you a new heart and a desire to be faithful. You will have only pure thoughts." (CEV). In Ezekiel 33:11, the LORD tells the prophet, "Say to them, 'As surely as I live,' declares the Sovereign Lord, 'I take no pleasure in the death of the wicked, but rather that they turn from their ways and live. Turn! Turn from your evil ways! Why will you die, people of Israel?'" (NIV). The LORD is also speaking to us, urging us to submit to the rewiring of our brains with new neural pathways that are committed to serving Him. As mentioned earlier, our brain can change, develop and grow by thoughts, actions, and intentions. The new neural pathways needed are described by Paul in Ephesians 4:1-3, "I, therefore, a prisoner for the Lord, urge you to walk in a manner worthy of the calling to which you have been called, with all humility and gentleness, with patience, bearing with one another in love, eager to maintain the unity of the Spirit in the bond of peace." (NASB)

I believe that as we become humble, gentle, and patient, our brains create new neural pathways. We now know through neuroscience that the brain produces neurons in response to such acts as kindness and empathy. These neurons wire together into neural super highways for humility, gentleness, patience, and bearing with one another. Over time, when this focus is frequently repeated, it becomes habitual (automatic). These neural pathways are pleasing to the LORD. Thus, spiritual transformation is an activity that occurs in the brain as it is molded by the Holy Spirit and nurtured by the Christian through deliberate and focused attention on Christ and Christlikeness. The new Christ-centered neural pathways replace old tendencies.

Paul admonishes us to present ourselves to God as living sacrifices and not to conform to the pattern of this world but to be transformed by the renewing of the mind. (Romans 12:1, 2). In other words, Paul is saying that our minds can be transformed when submitted to God, and He will facilitate the rewiring of our brains. The LORD (the Potter), who created the brain is best suited to remold our neuroplasticity to form new neural pathways that have a God-pleasing mindset. He can replace our evil thoughts with Godly thoughts. God created the brain in such a way that it can be rewired repeatedly. With repetitive God-pleasing messages, new Christlike neural pathways will dominate, and old habits and practices will diminish.

The Bible declares that the LORD will keep us in perfect peace when our minds are stayed on Him (Isa 26:3). When we decide to follow the LORD and keep our minds on Him, brain plasticity produces networks of neural pathways triggered by that decision. Repetition of the decision results in permanent neural networks that control our actions, attitudes, and feelings that define our personality and disposition. Maybe the LORD had human neuroplastic capacity in mind when, through Jeremiah, He commanded: "So now, correct your ways and deeds, and obey the voice of the LORD your God, so that He might relent of the disaster He has pronounced against you" (Jeremiah 26:13 Berean Study Bible). Is it possible for us to correct our "ways and deeds"? Jesus said that we cannot correct ourselves without Him. In John 15:5, Jesus said, "I am the vine, you are the branches. He who abides in Me, and I in him, bears much fruit; for without Me you can do nothing." (NKV). To abide in Christ, we must allow the Holy Spirit to dwell within us and transform us. He will plant within us the desire to be pleasing to the LORD. Such practice produces a network of new neural pathways dedicated to pleasing the LORD. If we allow it, this new network of neural pathways will override or supplant the old neural pathways that have propensities to do evil. We become a new creation when we abide in Christ (2 Corinthians 5:17).

When Joshua asked the Children of Israel to choose whom they would serve (Joshua 24:15), he was invoking neuroplasticity. Choosing to follow the LORD requires the creation of new neural pathways of faith and faithfulness. Faithfulness replaces the old self. The LORD asked the people to turn from their evil ways because He wanted them to turn to Him. He created our brain with the capacity for neuroplasticity so that it can be rewired for right-doing (righteousness) when given in total and complete submission to Him.

Figure 3. A Visual Summary of Spiritual Transformation that occurs when our thoughts are singularly focused on the LORD Jesus Christ and His love for us. This produces new neural highways while pruning away the old nature in us.

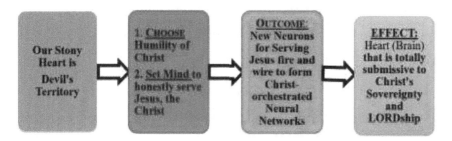

Chapter 5
Neuroscience Of New Creation In Christ

2 Corinthians 5:17 (NKJV)

The brain imprints and encodes every perceived message, environment, experience, emotion, and attitude, as well as what you say to yourself (self-talk). As a man thinks, so is he, Proverbs 23:7). Thus, the brain is constantly restructuring itself through all our experiences. Our experiences are the fuel for this restructuring. Everything we see, perceive, feel, experience, sense, do, and think changes the architecture of our brains. Spiritual transformation is directing our experiences to those that please the LORD as prompted by the Holy Spirit, without resistance. This process creates neural pathways dove-tailed to please God. Human beings are born wired to sin. "For I was born a sinner--yes, from the moment my mother conceived me." (Psalm-51:5, NLT). "Therefore, the LORD commands all people everywhere to repent (Acts 17:30, ESV). Repentance rewires the brain to follow the LORD with the Holy Spirit as the catalyst. Neuroscience has shown that our thoughts and self-talk play a critical role in the wiring and rewiring of the brain. We understand from neuroscience that our deliberate and conscious intention to take a course of action increases the brain's ability to produce neural pathways in response. "I am fearfully and wonderfully made" (Psalm 139:14 NIV). I believe that when a person makes a conscious intention to follow the LORD, the brain produces neural pathways for doing that which pleases God, prompted by the Holy

Spirit. New God-pleasing neural pathways are formed and gradually replace old ones. Every person has the capacity and potential to produce God-pleasing neural pathways, just like a biological reaction "A" that has a potential to proceed to converted to "B". The process requires an enzyme (catalyst), namely, the Holy Spirit. "Can an Ethiopian change his skin or a leopard its spots? Neither can you do good who are accustomed to doing evil" Spiritual transformation is the work of the Holy Spirit molding our neuroplastic brains into God-pleasing human beings as they find deliverance in Christ,

Our brain is like a group of muscles: if you exercise one group of muscles, it gets stronger and stronger, and if you neglect exercise, it becomes weaker and weaker. Neuroplasticity studies indicate that when we talk in negatives, we are making that part of the brain more and more active, so these negative pathways become stronger. As a result, these unhelpful and negative pathways are more easily triggered and help us to think in the negatives. 2 Corinthians 5:17 is saying that when our thoughts and inclinations are Christ-centered, we become wired with new neural pathways. The old ones are pruned and displaced. The brain grows new neurons and rewires itself into a Christ-centered neural network that is reinforced by repetition. From the neuroscience point of view, if anyone rewires the brain with Christlikeness, that person becomes a new creation and has replaced the old ungodlike habits and practices. Thus, whatever we learned in the past should not count because the brain is constantly changing and overriding the past. Spiritual transformation is rewiring the brain to practice Christlikeness. This entire process is facilitated by the Holy Spirit as the catalyst. Because of God's mercy, regardless of our past, it is possible to induce our brain to follow Christ by accepting His Lordship in our lives through repetitive practices and commitment to serve Him. This creates engraved Christ-pleasing neural networks. We become a new creation in Him.

Equally, the less we use the negative pathway, the weaker and weaker it becomes and the harder it is to revert. The good news is that we can simply and quickly change the way our brain is wired, howbeit, not on our own strength. The LORD says "For my yoke is easy to bear, and the burden I give you is light" (Matthew 11:30, NIV). In John 15:5, Jesus said: "I am the vine; you are the branches. If you remain in me and I in you, you will bear much fruit; apart from me you can do nothing." (NIV) A branch does not have its own identity. It cannot exist without being attached to the tree.

Focussing on Christ and Inclination to Serve Him Creates God-Fearing Neural Highways

Paul says: for all of us who were baptized into Christ have clothed ourselves with Christ. Therefore there is neither Jew nor Gentile, neither slave nor free, nor is there male and female, for you are all one in Christ Jesus. In Gal. 3:27,28, Paul is describing a spiritually transformed person. A transformed person does not treat anyone differently because of race, color, tribe, or caste. No one can claim spiritual superiority. Christ died to make this possible. If we treat people differently because of race, color, tribe, or caste, we are spiritually deformed and in need of transformation through Christ-Pleasing Neural Pathways via complete surrender and total focus on the LORD Jesus Christ. A new creation is clothed with Christ and manifested by how we talk about and treat others. The brain records all that we think and say in public, private, or in self-talk. The brain produces neural pathways for every one of them, without exception. The prominent thoughts and self-talks create neural super highways that manifest themselves in behavior, etc., and how we treat others. Figure 4 outlines the proposed path to a transformed Christian.

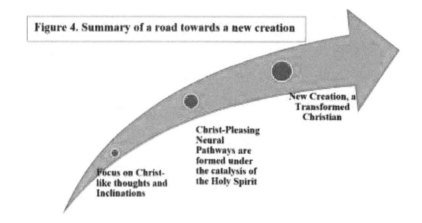

Chapter 6
The Neuroscience Of Mindset

Every person holds a set of attitudes or beliefs, collectively called mindset. Mindset is essentially a person's outlook and beliefs about oneself. Our mindset is crucially important and significant because our beliefs and attitudes impact everything we do, feel, think and respond to experience. Moreover, our mindset has a great influence on our perceptions and how we navigate the world around us. Some mindsets help us improve our wellbeing and succeed in life, while others, unfortunately, hurt our ability to succeed. Carol Dweck (2008), a professor of psychology at Stanford University, has conducted numerous studies on mindset. She has divided mindsets into two types: fixed and growth mindsets.

Fixed mindset. Dr. C.S. Dweck (2007) observed that when children believe that intelligence is fixed, they naturally avoid challenges, obstacles as well as setbacks. Their desire is to give the impression that they are smart. People with fixed mindset view challenging traits as naturally stable and unchangeable and do very little or nothing to change the circumstances. Adults with fixed mindsets make such statements as "This is who I am," which leaves no opportunity for growth or improvement. The phenomenon of neuroplasticity rejects the fixed mindset.

Growth Mindset. A person with a growth mindset understands that intelligence, abilities, and talents are learnable traits and are capable of improvement through determination.

To a person with a growth mindset, intelligence is a skill that can be developed, and a useful skill is the outcome of that effort, concentration, and focus. Failure is viewed and interpreted as a normal event in the process of learning. People with a growth mindset focus on (a) the task at hand, (b) the skill to be learned, and (c) the expected result. Difficulties are inevitable. When these difficulties arise, the growth-mindset person will be more resilient with perseverance and persistence (Dweck, C. S. (2008). Growth-mindset individuals perceive task setbacks as a necessary part of the learning process. Learners with a growth mindset are prone to embrace life-long learning and cherish incremental personal growth. Studies show that students taught a growth mindset have increased motivation and better achievement (Blackwell et al. 2007). This has implications for evangelism, community outreach, and educational enterprise.

Neuroplasticity makes growth mindset possible. When people are trained in growth mindset, new neurons are produced. These neurons wire together into growth-mindset neural networks. Like muscles, these growth-mindset neural pathways strengthen with repetition with the potential to become habits. Research shows that learners with a growth mindset exhibit efficiency in error-monitoring and are receptive to corrective feedback (Ng, B. 2018).

Growth Mindset and Spiritual Transformation

I propose that a Godly-growth mindset sets itself for spiritual transformation. It is this process of continuous adaptation that allows us to learn throughout our lifespan. As a malleable organ, our brain can be modified. As learning and experiences occur, neurons form neural pathways. If these neural pathways are Christ-centred, they are "planted in the house of the LORD, and will flourish in the courts of our God at any age. They will still bear fruit in old age; they will stay fresh and green." Psalm 92: 13, 14 (NIV). The brain of a Godly-growth-mindset person produces neurons that fire together and rewire to produce Christ-centered circuits of neural networks. At the same time, the brain disconnects existing ones that are "enmity with God." (James 4:4).

Growth mindset theory should be taught in schools, churches, and communities. Neural pathways of selfishness, egocentrism, and pride are eliminated or marginalized into oblivion and replaced by God-like thoughts, inclinations, and behavior. As mentioned earlier, neurons join to form neural pathways as one learns and experiences the environment. There is a neural pathway for whatever is learned or experienced.

Growth Mindset and Sanctification

Neuroplasticity is a life-time process. Neural pathways are formed, rewired, reorganized, or pruned away constantly for life. Sanctification is also a life-time process. I propose that Sanctification involves Holy-Spirit catalyzed neural wiring, rewiring, and reorganization of neural networks in response to our continual walk with the LORD Jesus Christ. Like muscles, these Christ-centered-growth-mindset neural networks get stronger with constant exercise as we become new creations in Christ. The feasibility of Sanctification is through our neuroplastic brains that God, Himself, created with the potential for a life-long journey towards Christ-likeness. This makes every human being equal before God in our capacity to undergo Sanctification through Holy-Spirit-induced neural super highways.

I venture to define Sanctification as a process of becoming Christ-like every day through taking micro-steps in Godliness. These small steps include (a) Christ-centered self-talk (chapter 10), (b) acts of kindness and attunement (chapter 12), (c) mindfulness (chapter 7), (d) growth mindset (chapter 6), mindsight (chapter 8), and (e) visualization of the glory of God (chapter 9) seen by Isaiah (Isa. 6:5) and Ezekiel (Eze1:1). I propose that these small steps facilitate the production of Christ-like neural pathways as the life-time of sanctification. Repetition of these steps strengthens the neural pathways that grow like muscles, as the brain produces neurons for each micro-step. These neurons fire together and wire together to form strong bundles of neural connections custom-made to please the LORD. Simultaneously, the old neural pathways are pruned away and displaced by the new God-pleasing neural pathways. This neuroplastic phenomenon is a life-time activity that must be exercised regularly like a muscle for it to maintain its strength and dominance over others. Figure 5 outlines my proposed process of Sanctification rooted in the adoption of a Christ-Seeking growth-mindset.

Figure 5. Developing the Growth-Mindset for Christ facilitates the Formation of Christ-induced Neuroplasticity, forming a network of Neural Pathways dedicated to serving the LORD.

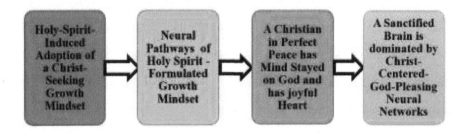

Chapter 7
The Neuroscience Of Mindfulness

"Mindfulness is a state of complete self-awareness without distraction or judgment. It is being in the moment with love and compassion, realizing the importance of the present and relating to it with peace and sympathy" (Douyon, 2019, page 81). It is being mindful that allows us to be aware of the physical and emotional changes taking place in our bodies and minds at any time. According to Dr. Douyon, a neurologist, being mindful is focused attention, allowing us to take inventory of our physical and mental being. It encourages our brains to attend to what is most important to us every moment and creates new nerve cells (neurons) and neural networks (Douyon, 2019). "Being mindful puts one on a path of awareness, compassion, creativity, and leadership, and in direct commands of neuroplasticity" *i.b.* page 84.

Thus, mindfulness is being aware of our thoughts, feelings, bodily sensations, and the environment without judgment or interpretation. It is described as the state of being fully cognizant of one's presence at any given moment. Being mindful is a skill that can be learned through practice with the following scientifically confirmed benefits: (a) working memory, (b) lowered anxiety, (c) improved visual attention and concentration, (d) improved emotional regulation, (e) improved mental resilience, (f) better sleep, (g) higher self-compassion, and (h) lower stress levels. (Tang et al. (2015) Moore, C. (2021), Widdett, R. (2014), Wheeler et al. (2017). As Christians, we know that there is a war between good and evil. Therefore, influencing and impacting neuroplasticity mandates that Christians be mindful of all positive and negative things that affect the well-being and functions of our brains. Being mindful is a very important exercise since it keeps us aware and cognizant of whatever has an influence on our brains while simultaneously changing the brain structure and function by building new networks and reorganizing or pruning the old ones.

Christian mindfulness is the awareness of God's active presence, and involvement in all that we do, say, and action at every moment of the day. In other words, our mind must be constantly and continually tuned to the LORD Jesus by listening to the promptings of the Holy Spirit as a still, small voice. Whenever the mind is turned away because of distractions, it must be tuned back to the LORD. This way, our minds will stay on the LORD, and we shall have perfect peace promised in Isa. 26:3. As stated earlier, spiritual transformation is a biological reaction. When one surrenders to God, the Holy Spirit is the catalyst in the formation of God-serving networks.

An individual must continually be aware of the fact that their brain is recording their views, thoughts, feelings, perceptions, and beliefs. The neuroplastic brain encodes every one of these as neural pathways that determine personality and disposition. The computer is encoded to go into default mode if not instructed otherwise. Doing evil is our default mode. David declared, "Behold, I was shaped in iniquity; and in sin did my mother conceive me (Psalm-51:5 NIV). The brain is naturally wired to do evil. It needs to be programmed to do right. Neuroplasticity makes this possible, and Christ-centered mindfulness makes this possible.

I believe that when the thief on the cross observed how Jesus was abused by the crowd and never said a word and was nailed to the cross without protest, his neural pathways were indelibly imprinted, and he declared: "And we indeed are suffering justly, for we are receiving what we deserve for our crimes; but this man has done nothing wrong." (Luke 23:41 NIV). Of interest is the fact that both thieves observed the treatment of Jesus, and yet it had opposite effects. Unfortunately, we spend very little time being mindful (our pondering time in the present) of what is going on at the moment. Instead, we spend a lot of time thinking about the past (often ruminating on painful things that happened) or being nostalgic about the pleasant past, or often worrying about the future. (Killingsworth & Gilbert (2010). Jesus said, "So do not worry about tomorrow, for tomorrow will care for itself.

Each day has enough trouble of its own" (https://bible.knowing-jesus.com/Matthew/6/34).

Spiritual Transformation is Intentional and Attentive

Neuroscientists have established that what we pay attention to and what we constantly expose ourselves to work synergistically to change our neural networks, which are manifest in our thinking and conscious awareness (Bosman, 2021). Scientists tell us that whatever we pay attention to grows. Anything we focus on dominates. Whatever occupies our time thrives. "Attention is like a spotlight, and what it illuminates streams into your mind and shapes your brain. Attention is also like a vacuum cleaner, sucking its contents into the brain. Directing attention skilfully is, therefore, a fundamental way to shape the brain – and one's life over time." (Wong, M.A. (2016). If we learn to consciously control our thoughts and behavior, it is possible to invoke self-directed neuroplasticity choices. (Hanson, R. (2013), Doidge, N. (2007), Amen, D. (1996), Begley, S. (2007) Schwartz & Begley (2003).

Again, anything that we pay attention to can change our brains and our lives. Therefore, choosing what to listen to should be done with great care, discretion, and caution. We can intentionally rewire our brains to create positive habits by deliberately focussing exclusively on the LORD Jesus Christ who stands at the door of our brains knocking for the Holy Spirit for facilitate spiritual transformation. This self-directed neuroplasticity is done primarily through active reflection. Because of what is called "experience-dependent neuroplasticity" and self-talk, neuroplasticity and whatever we hold in attention have a special power to change our brains. Thus, deliberately steering our attention in this way is both a decision to make and a skill to be learned. (Hanson, R. (2015).

Focus Leads to Spiritual Transformation

Be clear on what you want so that the brain is aware of the course of action. There is a need for brain single-mindedness to produce a neural imprint of what you want. If your singular focus is to please the LORD, the Holy Spirit induces the brain to generate neural pathways that manifest those things that please the LORD. When this is repeated continually, it supersedes and displaces sinful tendencies by suffocating them or starving them to death. The outcome is the fruit of the Spirit (love, joy, peace, longsuffering, gentleness, goodness, faith, meekness, temperance - Galatians 5:22). This becomes indelibly encoded into the brain neural networks and becomes our modus operandi of daily living. When this is repeated often, the brain wires it into a permanent way of life. This is spiritual transformation.

Figure 6 is a graphic illustration of the interconnectedness of mindfulness of God's presence, focussing on pleasing Him with intentionality, and attentiveness, culminating in a Christlikeness life.

Figure 6. Christ-Like Mindfulness: Sensing God's Presence Continually

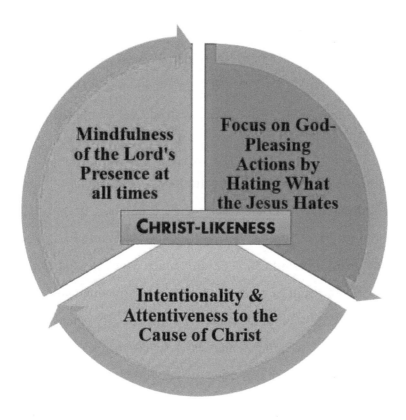

Chapter 8
The Neuroscience Of Mindsight

The Brain Filtering System

Our brains have a complicated filtering system that allows limited information that enters our conscious awareness and perception. Our brains tend to permit awareness of only what fits our existing beliefs, ideas, biases, and schemas (Frith & Jones, 2021). This short-cutting, unfortunately, fortifies and buttresses what we already believe and know or what we think we already know and experienced in the past. Thus, we write people off or marginalize what they are saying. Agarwal, P. (2020). This could be acceptable in an ideal world where the only filtering purpose is to protect us from total neural overload in everyday life.

Unfortunately, most of us are trapped in destructive thinking patterns, restrictive biases, obstructive habits, or impeding unfavorable emotional twists. Also referred to as "internal dialogue," "the voice inside your head," or an "inner voice," your internal monologue is the result of certain brain mechanisms that cause you to "hear" yourself talk in your head without actually speaking and forming sounds. The inner voice is a usual, often unnoticeable experience associated with activities such as verbal thinking and silent reading. After sin entere the earth, human beings developed the capacity to learn right and wrong. This learned behavior is stored in the conscience that serves as the inner voice. Consequently, whenever a person does something sinful, the inner voice tells the person that what he/she is planning is sinful. Then the person chooses to listen to or neglect the warning. Repetitive disobedience to the inner voice eventually silences it. Then the person does not feel guilty when lying, stealing, cheating, harboring hatred, keeping grudges, or revenging. The LORD gave humanity the power of choice which included choosing right or wrong. Neuroscientists have discovered that the power of the mind can actually change the structure and function of the brain through a process that Daniel Siegel calls *mindsight* and the 7th *sense*.

Mindsight, the 7th Sense

Daniel Siegel, MD, clinical professor of psychiatry at UCLA School of Medicine, has developed the theory of mindsight as an empowering action-oriented way of describing the power of the mind to actually change the structure and function of the brain. Mindsight describes the human capacity to perceive one's mind and that of others. It is mindsight that enables us to understand our inner lives clearly and integrate the brain while enhancing our ability to relate to other people. Siegel says that mindsight is like a reflective lens and focused attention that allows us to perceive the internal working of our own minds. Mindsight is our ability to look within ourselves, perceive our minds, and reflect on our experiences. This is an essential process, and Dr. Siegel refers to mindsight as the seventh sense. Mindsight is a learnable skill and can be cultivated through practical steps. Neuroscience studies show that developing the skill for mindsight actually changes the physical structure of the brain by stimulating the brain to grow important new mindsight-directed connections. Therefore, how we focus our attention shapes the structure and function of the brain. These studies further demonstrate that developing reflective skills of mindsight activates the very brain circuits that develop resilience, well-being, empathy, and compassion.

All human beings have the autopilot of ingrained behavior, attitudes, habitual traits, and responses. Siegel declares that *mindset* helps get us off autopilot, while *mindsight* allows us to identify and control the emotions we are experiencing. Thus, mindsight is not just being present every moment, but it includes being present to allow monitoring of what is actually going on both inside and outside. In addition, mindsight takes steps to modify what is happening. Mindsight can help us learn to focus attention on the mind itself to transform the connections in the brain, to move the brain to a more integrated, harmonious way of functioning. Thus, mindsight is the way we can focus attention on the nature of the internal world. It encompasses how we focus awareness on ourselves, our thoughts, and our feelings. Mindsight is also how we can focus on the internal world of someone else. It is how we have insight into ourselves and empathy for others. Mindsight is our communication with others and with ourselves. This communication helps us reflect on who we really are and what is going on inside us. This self-introspection has an impact on self-talk and focuses on spiritual transformation.

I believe that when Jesus, the LORD, says to us, "abide in Me and I in you" in John 15:4, He is telling us to have a mindsight for ourselves and a mindsight for others focussed on Him continually. For a Christian, mindsight should be controlled by the Holy Spirit. Such a mindsight is focused on Jesus and stimulates our brain to produce neural pathways that make us perceive His presence and nearness in our thinking, feelings, self-talk, and actions. When we are tempted to hurt other people, the new mindsight neurons overrule, and we bless them. When tempted to hate someone, the new mindsight neural pathways overrule, and we love the person. This is abiding in Christ. 'The high and lofty one who lives in eternity, the Holy One (Jesus Christ), says this: "I live in the high holy place and with those whose spirits are contrite and humble. I restore the crushed spirit of the humble and revive the courage of those with repentant hearts". (Isa. 57:15, NIV). A Christian with a mindsight focused on Jesus Christ becomes His dwelling place.

Figure 7 summarizes the proposition that the autopilot of negativism is interrupted by repentance. I propose that when repentance is followed by a continual focus on Christ Jesus (Christ-centered mindsight), the old autopilot is switched off by a Christ-

Figure 7. Christ-Centered Mindsight is God Pleasing

serving network of neural pathways.

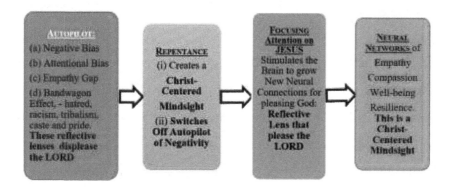

Chapter 9
The Neuroscience Of Visualization (Mental Rehearsal)

Visualization is a process of forming mental images. It is detailed imagination or mental rehearsal. Visualization is a powerful mind's eye that pictures oneself executing whatever it is one is planning to accomplish. In an effort to determine the impact of visualization, Alan Richardson, an Australian researcher, designed an experiment on basketball trainees. He divided the trainees into 3 groups for a 20-day training period.

Group #1 practiced free throws for 20 minutes every day between the first and last day of the 20-day training.

Group #2 was instructed to do nothing between the first and last day

Group #3 players were told to visualize and picture themselves making shots, missing free throws, and practicing correcting their shots every day between the first and last day. All players practiced making free throws on the first and last days of the 20-day training period. The results were amazing. The Group (#3) that did nothing but create mental rehearsals or just visualize making free throws performed almost as well as the Group (#1) that practiced every day for 20 minutes. The group that was instructed to do nothing did not show any improvement whatsoever. This study shows that visualization produces new neurons for an activity, even though unaccompanied by actual physical work.

Thus, it is possible to condition and strengthen the brain and its neural pathways by visualization, just as muscles and cardiovascular systems can. Visualization (mental rehearsal) generates new neurons that fire in unison and wire together to become neural pathways. If the mental rehearsal is Christ-centered, the new neurons fire and wire together to love other people as Jesus loves us. Repetition of these mental rehearsals solidifies into neural pathways that become neural networks of love, and our self-talk will now be, 'Let me love everyone, for love is from God.' (1 John 4:7).

Practicing Visualization

Peter Hollins (www.petehollins.com) proposed five steps for the concentrated immersion in our thoughts and imagination for visualization. The steps have been adapted is this book.

The first step is <u>relaxation</u>, a state of being free from tension and anxiety.

The second step is <u>imagining the environment</u> accomplished by building a mental picture of each person that is a source of your hatred, grudge, and spirit of revenge. Visualize a scenario of complete obedience to the LORD in our relationship with that person.

The third step is a <u>view yourself</u> in the third person. Picture yourself loving people the way Jesus loves us while keeping in mind John 3:16 and Galatians 5:14. Have a mental picture of others treating you the way that you want to be treated. This helps our neuroplastic brain to feel the impact of obeying God's command in Leviticus 19:18 " ---- but love your neighbor as yourself. I am the LORD." (NLT)

The fourth step is <u>viewing it in the first person</u>. Do an intensive imagining of yourself loving people (images of your God), not keeping grudges, and eliminating the desire to revenge. Then ascertain how your senses and emotions react and feel while imagining the three above. Remember that there are people in your circle who know about your hatred, grudges, and desire to revenge. How will you deal with them? Paul had difficulty convincing people that he was a changed man.

Step five is <u>coming back to reality</u>. Slowly re-emerge from your visualization into the real world, ready to take on the challenges of the real world. Your armor is an army of strong neural networks for obedience to God. In going through each of the five steps, you will follow the admonition: "Trust in the Lord with all your heart and do not lean on your understanding. In all your ways, acknowledge him, and he will make straight your paths." Proverbs 3:5,6(ESV).

I am convinced that a sincere visualization (beholding) of the goodness, love, and grace of the LORD Jesus Christ will create Godly neural pathways. When this visualization is repeated many times, the neural pathways are engraved to become a habit. This, I believe, is having the mind of Christ. "By beholding, we become changed into His likeness, for we are complete in Him.", (E.G., White, 1897).

Mirror Neurons

Neuroscientists have made an amazing discovery in our brains called the *mirror neurons*. A mirror neuron is a neuron that fires both when an individual act and when the person observes the same action performed by another. Thus, the neuron "mirrors" the behavior of another, as though the observer were the one acting. The unique characteristic of mirror neurons is that visualizing an action stimulates the same part of the brain as physically performing it. In contrast, other neurons are associated with either execution or observation, but not both. Mirror neurons do both. I believe that Christ-pleasing mirror neurons are produced by beholding/visualization of God's presence continually. When this is repeated often enough, permanent mirror-neuronal pathways become engraved as we are changed into His likeness. In doing so, the brain responds by producing **mirror neurons** which combine to form a strong neural network centered around the glorification of the LORD, His Majesty, and His love for us.

Gallese, at the University of Tubingen, Germany, stated, that it seems we're wired to see other people as similar to us rather than different and we identify the person we're facing as someone like ourselves." (Gallese et al., 2009). Further, researchers have discovered that some mirror neurons don't just care about what another individual is doing, but they also care about how far away they're doing it and, more importantly, whether there's potential for interaction. Because mirror neurons respond equally when we perform an action and when we witness someone else perform the same action, I believe that when the LORD commands us to follow Him, He knows that He created mirror neurons that make this possible. As we learn of Him in the Bible, mirror neurons are capable of making us do the works that He did.

Spiritual Transformation and Mirror Neurons

Apostle Paul might have been referring to mirror neurons when he wrote: "And we all, with unveiled face, continually seeing as in a mirror the glory of the Lord, are progressively being transformed into His image from [one degree of] glory to [even more] glory, which comes from the Lord, [who is] the Spirit." (2 Corinthians 3:18, AMP). By imitating the LORD Jesus continually, our mirror neurons become strong pathways, and we are progressively transformed into the His likeness.

When Paul says, "follow me, and I follow Christ," he admonishes us to visualize him as he followed Jesus. This action is facilitated by mirror neurons that respond equally when we perform an action and when we witness someone else perform the same action. I believe that when we have mental rehearsals of the goodness of the LORD and His love for us, the brain produces mirror neurons that make us mindful and aware of His presence in our lives. I further believe that these mirror neurons enable us to follow Jesus by following His example since they are a mechanism for action-understanding, imitation-learning, and the simulation of other people's behavior. Hence Paul says, "Imitate me, just as I also imitate Christ" (I Corinthians 11:1, NKJV). This is corroborated by neuroscientists' discovery of mirror neurons which fire when we observe an action performed by another person. I postulate that when Christians behold, visualize, and imitate the LORD Jesus Christ, our brains produce mirror neurons that form mirror-neuronal networks, helping us to experience His continual presence (See Figure 8).

Telling our brains what is important

The brain has a bundle of neurons called the Reticular Activating System (RAS) designed to filter out unnecessary information so that important things get through. Frequent visualization allows stimulates the RAS to bring one's goals into conscious awareness. Scientists tell us that we can help RAS know what is important by using visualization to bring our goals to the forefront of conscious awareness. I believe that if our goal is spiritual transformation, visualization of the Glory of the LORD will bring this to our conscious awareness. A recommendation is that we should remind our RAS what is important by frequent and daily visualization.

Figure 8. Visualization And Imitations Induce the Brain To Produce Mirror Neurons that facilitate and Help Us Learn Through Mimicry And Imitation, directing the Reticular Activating System

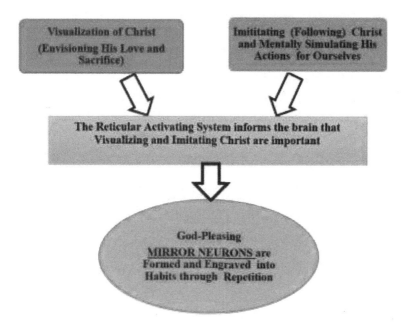

Chapter 10
Neuroplasticity And Self-Talk

As a person thinks in his/her heart, so is he/she. (Proverbs 23:7)

"Self-talk is the way you talk to yourself or your inner voice. You might not be aware that you're doing it, but you almost certainly are. This inner voice combines conscious thoughts with established beliefs and biases to create an internal monologue throughout the day". (https://www.healthdirect.gov.au › self-talk). Self-talk is everything that we verbalize, audibly or silently, expressing what we feel about ourselves or others. Self-talk is where our beliefs, opinions, attitudes, and thoughts about ourselves, our lives, and our associations with other people and things are stored.

The brain records all forms of self-talk. Self-talk directs neuroplasticity. Since self-talk is the prime programmer telling our brains what programs to wire in, therefore, what we say when we talk to ourselves about anything (especially about other people) is critical (Helmstetter, 2013). Unfortunately, self-talk that is negative, is harmful to oneself, and especially harmful to other people. Notwithstanding, negative self-talk is part of the human experience. As Christians, we need the Holy Spirit to help us control negative self-talk. Self-talk that does not bring joy, happiness, and peace of mind, is creating negative neural pathways. When we challenge negative thoughts, we are cultivating self-awareness. Positive self-talk, on the other hand, is a conscious and deliberate effort to positively self-direct what we say aloud, think, or otherwise express regarding our feelings about anyone or anything.

I believe that positive self-talk can easily be accomplished when our minds stay focused on the LORD (Isa. 26:3). If our self-talk is Christ-centered, then, both our thoughts and what we say about ourselves and others, allow the Holy Spirit to rewire our brains so that we want to do to others what we want others to do to us.

Not only is positive self-talk healthy for the individual, but it is also beneficial to others. With the discovery of neuroplasticity, we know that self-talk plays a pivotal role in creating a healthy outlook in life. Thus, self-talk is the fulcrum of neuroplasticity. The brain gets instructions from self-talk to wire lessons learned into neural pathways. Repeated instructions become semi-permanent neural networks (habits) that we perform without thinking. Thus, Christ-centered, habitually positive self-talk creates neural pathways of positivity.

In his book, *The Power of Neuroplasticity- What to say when you talk to yourself,* Helmstetter, asserts that self-talk is the prime programmer telling our brains what to wire in. He goes on to state that self-talk forms our beliefs, sets up our personalities, and determines whom we become. Dr. Helmstetter further states that anything we say out loud or think silently to ourselves, that expresses our feelings about ourselves, others, or anything else, is self-talk. Whether self-talk is positive or negative, it is being wired by our brains in its entirety, (Helmstetter, 2013). When the LORD admonishes the wicked to turn from his evil ways and live (Ezekiel 33:11 KJV), I believe that the Ceator is reminding us, that He created us to glorify Him, gave us brains that can be molded for good or evil, and He is encouraging us to reject evil. God the Holy Spirit can rewire our brains, no matter how far we have sunk in sin, if we are willing to turn from evil and invite Him into our lives. Christ-centered self-talk opens our minds to the Holy Spirit to rewire our neuroplastic brains in ways that glorify the LORD Jesus and Him alone. I believe that God-centered self-talk, that demonstrates that we choose Christ and reject Satan, triggers the rewiring of the brain, which was custom-designed by the Creator to withstand the wiles of the devil (see Genesis 3:15). Every repetition of edifying self-talk strengthens the neural pathways into permanent super neural networks that replace the old evil tendencies with a singular goal to please the LORD, come what may.

The things we repeatedly say to ourselves, how we say them, and the stories we tell ourselves about ourselves and others in our lives are wired into neural pathways (Gainsburg & Kross (2020). Because of neuroplasticity, our brains change, develop, and grow and self-talk plays a major role. Therefore, if our self-talk is continually focused on Jesus (loving other people, helping anyone in need, being honest, truthful, and perceiving the image of God in all people), our lives will be Christlike.

Self-talk repetition becomes wired by the brain into habits. I propose that this is the transformation and mind renewal described by Paul in Romans 12:2. "Do not conform any longer to the pattern of this world, but be transformed by the renewing of your mind." This Christlikeness results in a new creature that has replaced the old nature with a structurally and functionally new brain.

David said, "Behold, I was shaped in iniquity; and in sin did my mother conceive me" (Psalm 51:5). Because of sin, our brain seem to be are wired to prefer amplifying negative experiences and perceptions over positive ones. Negativity seems to be our default response. This is because we naturally and continuously ruminate over our failures and what went wrong in the past. These negative thoughts and emotions become part of our internal dialogue (Bosman, M. (2012). During King Saul's self-talk, he said: "They have credited David with tens of thousands, but me with only thousands. What more can he get but the kingdom?" (I Samuel 18:8, NIV). As a result of this self-talk, King Saul spent the rest of his tenure as King of Israel seeking to kill David. Saul had become a different person since his anointment as king. After Samuel anointed him king of Israel, the Bible says "So it was, when he had turned his back to go from Samuel, that God gave him another heart; and all those signs came to pass that day." 1 Sam.10:9 (NKJV). The signs included prophesying. I believe God instantly transformed Saul's neural pathways to prepare him to serve the LORD with prophetic abilities. Unfortunately, pride, envy and desire to please human beings rather than follow God's instructions, displaced the God-serving neural pathways. Consequently, the Spirit of the LORD left him.

Researchers have found that perpetual self-criticism produces measurable negative changes in the brain, which also correlates with low levels of happiness and diminished life satisfaction. (Kyeong et al (2020) This has to do with adjustment of functional connectivity by self-talk.

Changing the way we unconsciously talk to ourselves makes a huge difference in our performance, since our self-talk influences cognition, emotion, feeling, action, and the well-being of our body

(Hardy, Oliver, and Tod, 2010). As mentioned earlier, this is self-directed neuroplasticity. By changing our language from negative to positive, from de-motivating to motivating, and from unhelpful to helpful, we can powerfully impact how we feel productively and ensure that our focus is on the 'right' things, and therefore increase our chances of success.

"For as he thinks in his heart, so is he" (Proverbs 23:7(NKJV). Our brains shape themselves based on whatever they see, hear, perceive, or are exposed to. Interestingly, this neuroplasticity occurs whether you state your intention verbally or mentally. The brain is unable to distinguish between an action and a thought. Thus, whether we say something verbally or in our mind, new neural pathways are formed, nonetheless. These become engraved as habits if repeated often. When the Bible says: "You will keep in perfect peace all who trust in you, all whose thoughts are fixed on you!" (Isa. 26:3 NLT), it is referring to the phenomenon that when your thoughts are focused on the LORD, the outcome is perfect peace. Jesus said: "…do not worry about your life, what you will eat or drink; or about your body, what you will wear," (Matthew 6:25 NIV), because focusing our thoughts on these things produces neural pathways of worry and if repeated often become habits that induce chronic stress, a precursor of diabetes, hypertension and so on.

Biblical Self-Talk

A deliberate, intentional and conscious effort to insert God and His Word in our thought processes is *Biblical Self-Talk*. New neurons focused on the LORD Jesus are formed according to the first law of neuroplasticity, which states that neurons that fire together, wire together. As stated before, the catalyst for Biblical self-talk is the Holy Spirit. Repetition of the focus on the LORD strengthens these Christ-centred neural pathways into habits. The outcome is described in Psalm 119:165 "Great peace have they which love thy law: and nothing shall offend them" (KJV). Again, self-talk allows us to reform our brains, devoid of external influences and perceptions. Thus how you talk to yourself is crucial.

I believe that self-talking (for example, repeating Leviticus 19:17 "You shall not hate your brother in your heart" AMP) in the third person could be very effective in creating neurons of love that fire together and wire together into neural pathways of loving obedience to the LORD. Through the natural process of pruning, the old feelings of harboring hatred are eliminated. This is a spiritual transformation.

As described earlier in this book, self-talk is the dialogue that occurs in our heads. They can impact our cognitive and physical performance along with our overall mental health and happiness. (Hatzigeorgiadis et al (2011) & Kim et al (2021). Positive self-talk is also about showing compassion, forgiveness, acceptance of other people and seeing mistakes as opportunities to learn and grow. A powerful form of self-directed neuroplasticity is deliberately choosing to practice positive self-talk that can enhance performance and mental health (Brzosko, M. (2019, & Surdek, S. (2018).

Growth Mindset and Self-Talk

Since our brain is malleable, we can generate new neurons and supplant old ones. "Therefore, if any man is in Christ, he is a new creature: old things are passed away; behold, all things have become new." 2 Corinthians 5:17 KJV). In spiritual transformation, the Holy Spirit-prompted growth-mindset neural pathways replace the old tendencies, practices, beliefs, attitudes. For example, if my constant self-talk is "Every human being is the image of God and the LORD commands me to love them," then the brain responds by producing neurons that see every person as someone deserving of my love. If my self-talk is "Do not tell lies," my malleable brain produces new neurons that supplant the old lying neurons. If my self-talk is "Do not steal," my plastic brain synthesizes new neurons that respect other people's property. When my self-talk is "Do not kill by character assassination," neurons of loving kindness replace the old that neurons of hatred. If my self-talk is "Do not covet," my malleable brain produces new neurons that rejoice and celebrate when others succeed, excel or prosper. The old self has been supplanted by a person whose growth mindset believes what Jesus says in John 13:34, 35: "A new command I give you: Love one another. As I have loved you, so you must love one another. By this everyone will know that you are my disciples if you love one another." (NIV)

When such self-talk is repeated over and over in the realm of growth mindset, permanent neural pathways for loving everyone, truthfulness, etc. become permanent fixtures of the brain. You have become a new creature in right-doing. When Paul says: "…work out your own salvation with fear and trembling; [13] for it is God who works in you both to will and to do for *His* good pleasure." (Philippians 2:12, 13 NKJV), he is, in fact telling us that when we humbly submit to God ("with fear and trembling and submission), The Holy Spirit will transform our minds (to will and to do of His good pleasure"). In other words, He will guide our self-talk to be like the mindset of Christ stated in Philippians 2:5-7, "In your relationships with one another, have the same mindset as Christ Jesus: Who, being in very nature God, did not consider equality with God something to be used to his advantage; rather, He made Himself nothing by taking the very nature of a servant, being made in human likeness." (NIV)

It was prior experience with God, time spent talking to Him, and developing a relationship of trust with Him that sustained many Old Testament faithful servants of God in the face of threats to their lives. In addition to praying, the Hebrew boys must have engaged in God-inspired self-talk individually and collectively to take their stand against King Nebuchadnezzar when facing a fiery furnace. It was self-talk to trust the LORD that made David fearless when facing Goliath. It was self-talk that made Stephen pray for the forgiveness of his killers. It was self-talk that Mordecai refused to bow to Haman, the Agagite, the enemy of the Jews when he passed by. Spiritual Transformation involves self-talk to believe, trust and depend on the LORD no matter what.

Chapter 11
Humility: A Source Of Christian Joy And The Opposite Of Pride

Humility is not thinking less of yourself,
it is thinking of yourself less
Rick Warren

Humble people have feelings or attitudes that they have no special importance that makes them better than others. They treat others as beings made in the image of God.

There is no respect for others without humility in one's self.
Henri Frederic Amiel.

Humility is consideration of how you truly and honestly feel about yourself. You can only appreciate others if you have a great sense of the fact that you were created in the image of God who loves everyone unconditionally. This great sense of self-worth makes you praise and encourage others. Humility is realizing that you and others are equally valuable as every human beings, created in the image of God.

Humility is nothing but the disappearance
of self in the vision that God is all.
Andrew Murray

Honest-humility is a basic personality trait representing "the tendency to be fair and genuine in dealing with others, in the sense of cooperation with others even when one might exploit others without suffering retaliation" (Ashton and Lee 2007, p. 156).

"Christ is the humility of God embodied in human nature; the Eternal Love humbling itself, clothing itself in the garb of meekness and gentleness" (Andrew Murray, Humility. Cold Tappan, NJ: Fleming H. Revell, page 17). In Philippians 2:5-8, Paul is underscoring this point. He says that we must admire the humility of Jesus and want it for ourselves. In the context of neuroscience, for this to happen, we must have our minds stayed on the LORD because we trust Him (Isa 26:3). Keeping our minds stayed on and attuned to what God commands produces new neural pathways for pleasing the LORD.

We need to understand what Jesus meant when he admonished us to humble ourselves. According to Greek scholars, Jesus used "tapeinos", which conveys the concept of having an accurate view of ourselves before God and others. (Colin Brown, 1967 & Tarrants, 2011). Humility, therefore, is having an accurate and realistic sense of who we are before God and others. On the contrary, pride is an exalted, puffed-up, and exaggerated sense of who we are concerning God and others. As Christians, we should constantly remind ourselves that all of us are God's creatures: small, finite, helplessly dependent, limited in intelligence and ability, sinful (Heb. 9:27). When we treat others unfairly and cruelly, it is because we do not have a clear and accurate view of God and ourselves. We must refuse to be preoccupied with ourselves and our importance and seek to love and serve others (Phil 2:3, 4). Based on our understanding of brain plasticity: by rejecting our preoccupation with ourselves and rejecting our own importance and self-preservation, we create neural pathways that reorient us from self-centeredness to other-centeredness. These engraved humility-based neural pathways manifest themselves in our serving and caring for others with a God-pleasing mindset. Thus spiritual transformation involves humility before God and others.

Prophet Elijah, a Hallmark of Humility

Humility is complete surrender to God. I believe that Elijah is an example of a human being who was so humble that the LORD took him to heaven in a glorious chariot of fire. In I Kings 17, we are introduced to Elijah, the Prophet. The LORD sends him to King Ahab to announce a 3-year drought because of Israel's wickedness. The LORD sends Elijah to a brook with water to drink and commands ravens to bring bread and flesh twice daily. Then the river dries up, but Elijah does not panic or complain to God about the disappearing source of water. He simply waits for orders from the LORD. He is then sent to a foreign country where the LORD had commanded a widow to take care of him. Jesus said: "I assure you that there were many widows in Israel in Elijah's time when the sky was shut for three and a half years and there was a severe famine throughout the land. Yet Elijah was not sent to any of them, but a widow in Zarephath in the region of Sidon." Luke 4:25, 26 NIV). Elijah obeyed God completely. He trusted God intrinsically.

Elijah's Complete Confidence in God.

In preparation for the sacrifice to his God to demonstrate the true God, he repaired the altar of the LORD and put wood and the sacrifice on top. He then ordered that 640 liters (168 gallons) of water be poured on the altar. He called upon the LORD to demonstrate His power. Fire from the LORD consumed the sacrifice, wood and all the water. Elijah had a special and personal relationship with the LORD, God of Heaven.

Humility and not Self-glorification and Fame:

When Elijah found out that he was going to heaven in a heavenly chariot of fire, he did not make any announcement about this uniquely momentous event. There was no fanfare whatsoever. There was no celebration and self-aggrandizement. The only person to witness this event in close proximity was Elisha. This is humility in great proportion and a hallmark of what we could be to please the LORD. Oh! To be like Elijah in humility and total dependence on the LORD.

Moses, another giant of humility

"Moses was a very humble man. He was more humble than any other man on earth." (Numbers 12:3, Easy-to-Read Version). "By faith Moses, when he became of age, refused to be called the son of Pharaoh's daughter, choosing rather to suffer affliction with the people of God than to enjoy the pleasures of sin for a time. He esteemed the reproach of Christ as greater riches than the treasures in Egypt, for he looked to the reward," (Hebrews 11:23-26, Modern English Version).

Elisha:

Naaman was the chief commander of the army of the king of Aram. He was a great man in the sight of his master and highly regarded because through him the Lord had given victory to Aram. He was a valiant soldier, but he had leprosy (2 Kings 5:1) This famous and gallant Naaman travelled to meet Elisha. To the average person, this would be a great and rare opportunity for Elisha to be famous. But not Elisha. He did not even come out of his house to greet his famous visitor. Instead, he sent a message for Naaman to go and wash seven times in the river. Notice the contrast in the two men. Proud Naaman thought dipping in the muddy Jordan was humiliating to a man of his station. But Elisha, by seeking no fame nor riches for himself, instead he sent a servant to meet his famous guest. This is humility.

As an example for us, the LORD Jesus made himself nothing by taking the very nature of a servantbeing made in human likeness and by sharing in human nature. "Think of yourselves the way Christ Jesus thought of himself. He had equal status with God but didn't think so much of himself that he had to cling to the advantages of that status no matter what. Not at all. When the time came, he set aside the privileges of deity and took on the status of a slave, became *human*! Having become human, he stayed human. It was an incredibly humbling process. He didn't claim special privileges. Instead, he lived a selfless, obedient life and then died a selfless, obedient death—and the worst kind of death at that—a crucifixion." (Philippians 2:5-8 The Message)

Humility alters God's Verdict: Ahab King of Israel

The Bible says that there was never a king of Israel who sold himself to do more evil in the eyes of the LORD than Ahab. (I Kings 21:25, NIV). However, when the LORD pronounced judgment on him, he humbled himself before God and the LORD responded by saying "Because he (Ahab) has humbled himself, I will not bring this disaster in his day, but I will bring it on his house in the days of his son." I Kings 21:29, NIV). Ahab's sincere humility altered God's pronounced verdict on him.

King Manasseh seduced Judah to do more evil than did the nations whom the Lord destroyed before the children of Israel (II Kings 21:9, KJV). "The Lord spoke to Manasseh and his people, but they paid no attention. So the Lord brought against them the army commanders of the king of Assyria, who took Manasseh prisoner, put a hook in his nose, bound him with bronze shackles, and took him to Babylon. In his distress, Manasseh sought the favour of the Lord his God and humbled himself greatly before the God of his ancestors. And when he prayed to Him, the Lord was moved by his entreaty (supplication) and listened to his plea; so He brought him back to Jerusalem and to his kingdom. Then Manasseh knew that the Lord is God". (2 Chronicles 33:10-13, NIV). Based on our understanding of neuroscience, by turning towards the LORD, Manasseh developed new neural pathways for fearing and serving the LORD. Humility triggered that formation of God-fearing neural pathways. The neural pathways that made him commit worse sins than all the kings before him were displaced by those embodied in serving his God in sincerity. Humility is a powerful, spiritually transforming tool.

King Nebuchadnezzar, in his address to the nations and peoples of every language, who lived in all the earth said "It is my pleasure to tell you about the miraculous signs and wonders that the Most High God has performed for me. How great are His signs, how mighty are His wonders! His Kingdom is an eternal kingdom; His dominion endures from generation to generation." (Daniel 4:2-3 NIV) He went on to say "I, Nebuchadnezzar, was at home in my palace, contented and prosperous." (Daniel 4: 4 NIV). Then came a dream whose meaning he did not know. Daniel interpreted the dream and gave the king the following advice: "Therefore, Your Majesty, be pleased to accept my advice: Renounce your sins by doing what is right, and your wickedness by being kind to the oppressed. It may be that then your prosperity will continue." (Daniel 4:27NIV). Unfortunately, the king did not heed Daniel's counsel. "Twelve months later, as the king was walking on the roof of the royal palace of Babylon," (in his self-talk) "he said, 'Is not this the great Babylon I have built as the royal residence, by my mighty power and for the glory of my majesty?' (Daniel 4:29, 30. NIV). Then a voice from heaven declared that his authority had been taken from him. He spent the next seven years in the wilderness until he recognized that the Most High is sovereign over all kingdoms on earth. (Daniel 4:31-32, NIV).

After seven years of humility, the king regained his sanity: "At the end of that time, I, Nebuchadnezzar, raised my eyes toward heaven, and my sanity was restored. Then I praised the Most High; I honored and glorified Him who lives forever. His dominion is an eternal dominion; his kingdom endures from generation to generation. All the peoples of the earth are regarded as nothing. He does as He pleases with the powers of heaven and the peoples of the earth. No one can hold back His hand or say to Him: "What have you done?" (Daniel 4:34-35, NIV).

There is clear evidence that while the king of Babylon was in the wilderness, his self-talk and mindsight changed and caused him to declare: "Now I, Nebuchadnezzar, praise and exalt and glorify the King of heaven, because everything He does is right and all his ways are just. And those who walk in pride He can humble." (Daniel 4:37 NIV). This is the last that we hear about Nebuchadnezzar, a humbled man praising, exalting, and glorifying the King of Heaven. Humility is spiritually transformative. The king renounced his sins by doing what is right. The king's humility created God-pleasing neural pathways and suppressed the old sinful and wicked ones. The king had previously declared "It is my pleasure to tell you about the miraculous signs and wonders that the Most High God has performed for me." (Daniel 4:1, 2 NIV). Unfortunately, King Nebuchadnezzar went back to his prideful spirit. In Daniel 4:30, he attributed his successes and prosperity to himself when he said, "Is not this the great Babylon I have built as the royal residence, by my mighty power and for the glory of my majesty?" (Daniel 4:30, NIV), which resulted in his seven years in the wilderness. There is an inherent danger in being successful, powerful, and prosperous. To whom do we attribute all these to? In Philippians, Paul says: "You must have the same attitude that Christ Jesus had. Though he was God He did not think of equality with God as something to cling to. Instead, he gave up his divine privileges. He took the humble position of a slave" (Phil. 2:5-7). "In your relationships with one another, have the same mindset as Christ Jesus." (NIV). Figure 9 is a proposal for allowing humility to displace pride in the life of a Christian.

Figure 9. Acceptance of the Lordship of Jesus Christ brings humility & pride disappears

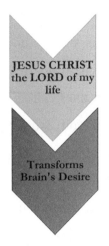

- JESUS CHRIST the LORD of my life
  - (1) Created us in His Image and implanted neuroplasticity which allows the Holy Spirit to tranform us back to His image, when we accept His Sovereignty in our lives.
  - (2) Emptied Himself for us because of His Love Us

- Transforms Brain's Desire
  - (3) Total Dependence on God
  - (4) Life characterized by humility
  - (5) Power Relinquishing
  - (6) Self-Abdication
  - (7) The same mind (attitude) in ourselves as demonstrated by Christ

The following are suggestions for humility:

- Have this same attitude in yourselves which was in Christ Jesus [Look to Him as your example of selfless humility].

- Do not merely look out for your own interests, but also the interests of others.

- "Do nothing out of selfish ambition or vain conceit. Rather, in humility value others above yourselves, not looking to your own interests but each of you to the interests of the others. In your relationships with one another, have the same mindset as Christ Jesus" (Philippians 2:3-5NIV).

- "Be completely humble and gentle; be patient, bearing with one another in love. Be kind and compassionate to one another, forgiving each other, just as in Christ God forgave you." (Ephesians 2:32 NIV)

- "When pride comes, then comes disgrace, but with humility comes wisdom." (Proverbs 11:2 NIV)

- The LORD, Jesus Christ created neuroplasticity to make possible self-abdication, just as He emptied Himself for us. (Phil. 2:7KJV). There is never an excuse for pride that puts others down.

Pride is the opposite of Humility

Pride will always be the longest distance between two people
Kushand Wizom

"Pride is at the root of all the atheism, theoretical or practical, on the earth; at the root of all the reluctance which there is to seek the favor of God; ..."

https://biblehub.com › commentaries › psalms › 10-4

C.S. Lewis (1996) has stated the following about pride:

> *Pride is dangerous because it hinders our intimacy with God and love for others. It is what separated Lucifer from God. It was pride that turned Lucifer into the Devil. Pride is a sin. It is the Devil's most effective and destructive tool. Pride is the cause of misery in every nation and every family. Pride is spiritual cancer. It eats up the possibility of love or commitment or even common sense. Chances are that most of us do not see the pride in our lives, but easy to see it in*

others. The more we have pride in ourselves, the more we dislike it in others. There can be no surer proof of confirmed pride than a belief that one is sufficiently humble. If you want to find out how proud you are, the easiest way is to ask yourself: How much do I dislike it when people snub me, refuse to notice me, shove their oar in or patronize or show off? Make no mistake about it: Pride is a great sin.

C. S. Lewis

Chapter 12
The Neuroscience Of Kindness

"Anyone who withholds kindness from a friend forsakes the fear of the Almighty" **(Job 6:14, NIV)**

Whenever we perform an act of kindness, there is a warm feeling of well-being that bathes over us. Neuroscientists discovered that acts of kindness release a brain chemical that contributes to a good-feeling mood and overall well-being. This kindness chemical is a hormone called oxytocin. Thus, because of the pivotal role of kindness in our well-being and that of others, the Creator installed a system in the brain that responds to acts of kindness to make both the giver and the receiver feel good. The kindness practice is so effective that it's being formally incorporated into some types of psychotherapy in medical practice. Oxytocin is tied to making us more trusting, more generous, and friendlier, while also lowering our blood pressure. Research suggests that just a simple act of caring touch seems to boost oxytocin release. Cuddling and hugging someone leads to higher levels of this hormone and results in a greater sense of well-being to the giver of kindness. "We all seek a path of happiness," says Dr. Waguih William IsHak, a professor of Psychiatry at Cedars-Sinai. He goes on to say "Practicing kindness towards others is what we know works in medical practice. However, acts of kindness must be repeated, He says, biochemically, one cannot live on the 3 to 4 minute oxytocin boost that comes from a single act of kindness. That is why kindness is most beneficial as a regularly repeated practice. The rewards of acts of kindness are many. They help us feel better and they likewise help those who receive them. When we do acts of kindness, we are building better selves and better communities at the same time. (https://bio.cedars-sinai.org/ishakw/index.html), a professor of psychiatry (/programs/psychiatry.html) at Cedars-Sinai).

Ballatt and Campling in their 2011 book, *Intelligent Kindness: Reforming the Culture of Healthcare*, summarise some of the evidence for the impact that kindness can have on our brains. They reiterate that individual acts of kindness release both endorphins and oxytocin and create new neural connections. The implications for such plasticity of the brain are that altruism (concern for the welfare of others) and kindness become self-authenticating. In other words, kindness can become a self-reinforcing habit requiring less and less effort to exercise. Indeed, data from functional magnetic resonance imaging (fMRI) scans show that even the act of imagining compassion and kindness activates the soothing and affiliation component of the emotional regulation system of the brain with the release of oxytocin (Ballatt and Campling, 2011).

A brain imaging study by the University of British Columbia showed that when we donate to charity, our brain's "pleasure center" lights up like the full moon on a clear night. Another study at the University of San Diego found that acts of kindness, generosity, and cooperation spread like wildfire to everyone nearby. University of Wisconsin neuroscientists' brain imaging studies show that this happens to be the same brain area that lights up like a Christmas tree when our "kindness & compassion barometer" are exercised. (Lutz, et al, 2008). Kindness can be strengthened like a muscle through constantly repeated use.

Cole-King and Gilbert have defined compassion (or kindness) as being sensitivity to the distress of others with an obligation to endeavor do something about it. The main point here is that, if we are to be kind, then not only do we need to be sensitive to the suffering of others, but we also need to make a constructive response in such circumstances. Kindness should be central to our engagement with others because it is central to healing." (Mathers, 2016).

Proverbs 11:17 (ESV) says, "Those who are kind benefit themselves, but the cruel bring ruin to themselves." "Therefore as God's chosen people, holy and dearly loved, clothe yourselves with compassion, kindness, humility, gentleness, and patience" (Colossians 3:12 NIV). Ballatt and Campling describe kindness as a virtuous cycle. Kindness directs attentiveness, which in turn enables attunement. Attunement builds trust. Attunement is how we react to other people's emotional needs and moods. A well-attuned individual responds with appropriate language and behavior based on the other person's emotional state. "When we attune to others, we allow our internal state to shift, to come to resonate with the inner world of another." (Daniel Siegel, https://cultureofempathy.com). Kindness has a positive impact on our brains. Neuroscience provides evidence that kindness builds a better world and a better person (Richard Hill, www.thescienceofpsychotherapy.com/neuroscience-kindness-and-compassion-doing-something-good-for-your-brain/).

Kindness is generational

The Perry Foundation (www.perry-foundation.org) donated millions of US dollars to grassroots charities that benefit children, families, and communities worldwide. The Chief Executive Officer and Founder of the Foundation, Mr. Tyler Perry was asked about the impetus to create the foundation, and he said that he has been given a lot. And the need is great. He said "my mother put it in me. I would wake up in the morning and get out of bed, and I'm stepping on someone who was sleeping on the floor. I'd ask, 'who's that?' And my mom would say 'They needed a place to stay. She did not have much, but what she did have, she shared. As I do this work, I'm always thinking of her." (AARP The Magazine- August/September 2022 aarp.org/magazine, page 85). The kindness and compassion demonstrated by Tyler's mother created neural pathways of kindness and altruism in him. Thousands of individuals and communities are beneficiaries. This kindness was transmitted from mother to son.

The Neuroscience of empathy with kindness

It is curious that people, in general, are frequently and routinely kind and generous, sacrificing largely without expecting anything in return. When I was a foreign student in Sierra Leone, I was walking down a street in a busy city and saw a taxi driver hit a 10-year child who was trying to cross the street. Bystanders just stood by and did nothing. Then a total stranger, a foreigner ran and picked up that child from the street and called for an ambulance. Neuroscientists at the University of California, Los Angeles (UCLA) have shown that if you feel physical pain, a part of your brain lights up. What is amazing is that if you see someone in physical pain, the same part of your brain lights up, which leads us to believe that the Creator wired human beings who have experienced pain (which is nearly everyone) to respond with empathy to others who are experiencing pain. The

Creator wired us for empathy. Therefore, He desires us to be empathetic toward others in need whose feelings and thoughts that may be different from our own. This is a complicated cognitive achievement. "Rejoice with those who rejoice, weep with those who weep." Romans 12:15 (NET). Scientists report that from a physiological point of view, kindness positively changes our brains. Not only does kindness induce the brain to produce oxytocin, the kindness hormone, it boosts serotonin and dopamine, which are neurotransmitters in the brain that give you feelings of satisfaction and well-being, and makes the pleasure/reward centers in the brain to fire (light up). In addition, endorphins, which are your body's natural pain killer, are also released. Figure 10 summarizes how acts of kindness trigger the production of brain chemicals that elicit feelings of satisfaction and well-being in the kind person.

Figure 10 summarizes the neuroscience of kindness in the context of the wellbeing of the giver of kindness.

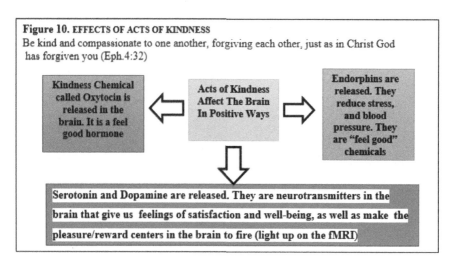

Chapter 13
The Neuroscience Of Habits

But solid food is for the mature, who because of practice have their senses trained to discern good and evil (Hebrews 5:14).

Habits are settled or regular tendencies or practices, especially ones that are hard to give up. Habits develop from repetition, which is the major tool of neuroplasticity. They are the brain's way of automating behaviors that do not require complex thought processes. For every habit, there is a well-trodden neural pathway. This enables the individual to devote their intellectual resources to challenging tasks. There is very little thinking involved in acting from habit. Neuroscientists tell us that repetition is essential. They state that repetitions allow a conscious skill to transfer into one's subconscious, freeing up working memory which allows for the learning of other skills. According to Schmelzer, repetition creates long-term memory by eliciting or enacting strong chemical interactions at the synapse of the neuron (where neurons connect to other neurons). Repetition creates the strongest learning—and most learning, whether implicit (like tying your shoes) or explicit (like memorizing multiplication tables) relies on repetition" (Schmelzer, 2021). A habit is doing something automatically (without thinking). New habits become automatic after repeating the behavior for an average of 66 days. (Lally, et.al 2009).

The prophet Daniel made it a habit to pray 3 times a day. This repetition reinforced his resolve to serve God no matter what. Even after the unalterable decree had been signed, which required all people in the kingdom to worship only King Darius, Daniel kept his habit of praying to the God of Heaven three times a day in his usual, and very visible location (Daniel 6:10,11). This was the result of Holy Spirit-directed neuroplasticity development which resulted in neural highways that had become a way of life for Daniel. Gabriel said to Daniel "for thou art greatly beloved" (Daniel 9:23, KJV). God-pleasing habits are the most powerful and effective mechanisms for the spiritual transformation of neural superhighways. Thus, spiritual transformation is allowing the Holy Spirit to facilitate the creation of neural superhighways for serving God and trusting Him implicitly.

A spiritually transformed person's mind must be repeatedly directed by the Holy Spirit to think and do that which pleases the LORD in order to establish a God-pleasing habit which produces neural superhighways. Again, in keeping with current thought in neuroscience, after 66 days of thinking and doing that which pleases the LORD, those practices may become nearly automatic and therefore done without thinking. However, because of the nature of the human heart, we are ever dependent on the Holy Spirit's promptings to do that which is Christlike.

Watch your thoughts, they become your words,
watch your words, they become your actions, watch
your actions, they become your habits; watch your
habits, they become your character; watch your character,
it becomes your destiny
Lao Tzu.

Example of Habits

I was born in a village in Zimbabwe. At a very early age boys were taught how to plough with oxen. The plough was pulled by oxen, making furrows. The furrows had to be straight. For the oxen to follow the furrow, a young boy would lead the oxen (Figure 11). After many repetitions, the oxen learn to follow the furrow without being led and would do this until the field was completed (Figure12). The oxen developed neural pathways for following the furrows after repetition. Animals are trained through repetition and that practice becomes a habit (Fig. 12).

Figure 11. Oxen are being trained to follow the plough path during ploughing. The boys are directing the oxen to plough along the path and stay on course.

Source: https://www.robertharding.com/preview/797-2850/gambia-farming-ploughing-oxen-drought/

Figure 12. These oxen have learned to pull the plough without a lead person. Source :https://www.robertharding.com/preview/1194-1157/zambia-farmer-ploughing-field-shangombo/

Repeated practice helped the oxen to develop the desired skill. This guided process is repeated many times until the oxen pull the plough without a leader as shown in Figure 12. Repetitions re-wire the brain to perform habitually and creates habitual behavior in the oxen. This process has created neural pathways for following the plough path without the oxen being led or directed. Through many repetitions and training, these oxen learned to pull the plough as taught, without a guide.

Figure 13. Steps for habit formation (Gardner, 2012) prompted by the Holy Spirit

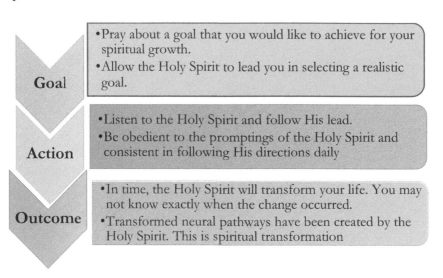

Thus, if we are led by the Holy Spirit and obedient to His instructions, we will practice Christlike actions on a regular basis, (i.e., 66 days and more), and permanent neural pathways for pleasing the LORD will become habits that control our thoughts, actions, and self-talk. "…bowels of mercy, kindness, humbleness of mind, meekness, longsuffering forbearing one another, and forgiving one another" become habitual. Colossians 3:12 (KJV).

Chapter 14
Wandering And Distractibility: Enemies Of Perfect Peace In Jesus

The Wandering Mind

Wandering is not paying attention to what you are thinking, or what is going on in the brain. Research shows that a wandering mind is a default mode (like a computer) but with deleterious health effects. A wandering mind may dwell on (i) what is going wrong in the world, (ii) bad events in the past, and (iii) what might happen in the future, among other things. It is a stimulus-independent thought process specializing in negativity, i.e. worrying about oneself to the point of unwellness.

A great deal of research has been done about cognitive and neural mind wandering. Kilingsworth and Gilbert (2020) reported from their research that the human mind is a wanderer if left alone. A wandering mind is an unhappy mind, since it tends to dwell on what is wrong with the world, and so on. The process of thinking about what is not happening is a cognitive practice that comes with a great emotional cost. According to Kilingsworth and Gilbert, human beings spend an inordinate amount of time thinking about what is not going on around them, contemplating events that happened in the past, might happen in the future, or will never happen at all. This stimulus-independent thought or mind-wandering appears to be the brain's default mode of operation (Raiche et al.2001, Christoff, et al, 2009; Buckner et al 2008).

Neuropsychologists conclude that a wandering mind steals our joy. Consequently, this unhappy wandering mind has deleterious health consequences. It can make a person sick. "A joyful heart is a good medicine, but a broken spirit dries up the bones." Proverbs 17:22 (ESV). I believe that having our minds stayed on the LORD Jesus Christ, who told us not to worry about tomorrow, gives us a joyful heart (good medicine). When the Bible says that God will keep the person in perfect peace whose mind is stayed on Him, it is admonishing us to stop our minds from wandering by concentrating on the LORD (see Isaiah 26:3-4).

Distractibility

Research has identified a major challenge in our society that is deleterious to our well-being in very important ways. It is distractibility, defined as "a fleeting attention span." (www.sciencedirect.com/topics/medicine-and-dentistry/distractibility). People who do not pay attention to what they are doing or thinking are often unhappy, which generally results in an unhealthy mind.

David Richardson, a psychologist/neuroscientist at the University of Wisconsin has proposed 4 pillars for a healthy mind:

1. The first pillar is **awareness.** Awareness includes the capacity to focus our attention to resist distraction. This includes a condition that psychologists and neuroscientists describe as meta-awareness. Meta awareness simply means to be conscious of and knowing what your mind is doing. Many people faithfully sit in church during a

sermon, yet, if asked what the preacher said, have no idea. But when a person recovers and pays attention, they experience a moment of meta-awareness. **Meta awareness** is crucial for genuine spiritual transformation to occur. Spiritual meta-awareness is re-focusing your mind on what is pleasing to the LORD, which is an essential part of spiritual transformation

2. The second pillar of a healthy mind is **connection,** which refers to those qualities that nurture harmonious interpersonal relationship qualities such as kindness, compassion, empathy, and having a positive outlook. According to research, it does not take much effort to start activating these latent qualities, and they can flourish and become stronger with repetition.

3. The third pillar of a healthy mind is **insight**. This is a narrative that we all have about ourselves. At the extreme end of the continuum, some people have a very negative narrative. They have negative self-beliefs, and they hold those beliefs to be a true description of who they are. This is a prescription for depression. A healthy mind entails changing our relationship to this narrative, not so much to change the narrative itself, but to change our relationship to the narrative in order to increase our well-being.

4. The fourth pillar of the healthy mind is **purpose.** By purpose, we mean that the person has a sense that their life is headed in a particular direction.

Studies show that distracted people are unhappy and unhealthy. (See Figures 14 & 15).

Figure 14. Harmful Effects of a Wandering Mind

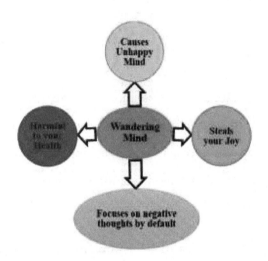

"It is Satan's constant study to keep the minds of men occupied with those things which will prevent them from obtaining the knowledge of God. He seeks to keep them dwelling upon what will darken the understanding and discourage the soul." (Ellen G. White, https://whiteestate.org/message/Character_of_God.asp). Thus, we conclude that Satan's primary objective is to keep the mind of a Christian wandering and distracted from Christ at home, in church, at work and when interacting with others. This results in thoughts and actions that are incongruent with that of a true and honest disciple of Christ.

Figure 15. Antidote for A Wandering or Distracted Mind with the Holy Spirit as the Agent

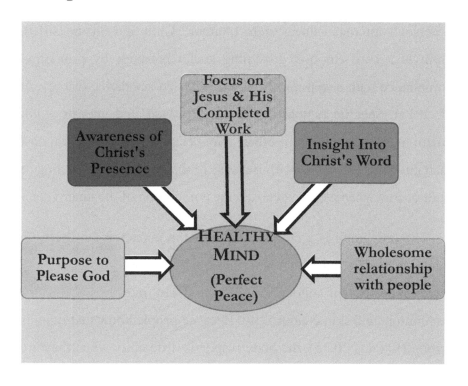

Chapter 15
Guilt: A Cause Of Chronic Stress In A Christian

Research shows that guilt occurs when our behavior conflicts with our conscience (inner voice). It is when we think that we have damaged our reputation or embarrassed others that we feel shame. Guilt is linked to a person's learned social standards that are established with a person's internal value system (Author? Guilt and Shame 2019). Guilt is a core emotion governing social behavior by promoting compliance with social norms or self-imposed standards. (Morey, et al, 2012). Specific regions of the brain activate and intensify guilt where harm to others (guilt-other) may occur as a greater social cost than guilt-self (Morey, 2012). Because of the inner voice, a feeling of guilt occurs whenever we overrule the admonition of the inner voice.

The polygraph is a clear example of the impact and role of the inner voice. The polygraph is a physiological recorder that assesses three indicators of autonomic arousal: heart rate/blood pressure, respiration, and skin conductivity. Because people know that lying is wrong (Exodus 20:15), the body responds differently when they are untruthful. The inner voice knows that the individual is lying and reacts to guilt physiologically. Therefore, the liar fails the polygraph test.

It is my conviction that the inmost being is the inner voice and conscience that God installed to remind us of the right thing to do. In Exodus 20:16, the LORD says "You must not tell lies about other people." (Easy-to-Read Version). When people lie, they know that lying is wrong, as evidenced by autonomic arousal: heart rate/blood pressure, respiration, skin conductivity, and temperature elevation, as measured by the polygraph. I believe that God installed these physiological reactions in the autonomic nervous system that functions without our conscience being aware, as a reminder that telling lies is not acceptable to Him, nor to the victim. The polygraph test assesses these indicators.

The Neuroscience of Guilt and Shame

There are neural networks for guilt and shame. I believe that the LORD designed us to feel guilt and shame to protect us against hurting other people and hurting ourselves. The discomfort that we feel encourages us to alter our behavior and treat others better in the future. Guilt is designed to make us kinder and more compassionate. Certain emotions are deeply painful. Guilt and shame fall in this category of human emotions.

As Christians, we are taught to become thoroughly acquainted with the 10 Commandments and their demands to love God and our fellowmen. Six of the 10 Commandments relate to how we should treat others. These commands include: honor thy father and thy mother; thou shall not steal; thou shall not kill; thou shall not covet; thou shall not lie; and thou shall not commit adultery (see Exodus 20). Christians also recite the LORD's prayer which reminds us: Forgive us as we forgive others. All these Bible passages become imprinted and engraved in every Christian's neural pathways. Whenever we break a Commandment, the brain's inner voice (conscience) broadcasts neuronal signals prompting us to seek forgiveness and restoration. Guilt is the consequence of disregarding the inner voice (the Holy Spirit urging us to choose good and not evil).

As Christians, we know right from wrong. We also know that Jesus admonishes us to "Love one another as I have loved you" (John 13:33, 34. NIV). In practice, however, we frequently break each of the Commandments, especially those that demonstrate love for others. However, the most painful transgressions are those disregarding the greatest commandment: "Master, which is the great commandment in the law? Jesus said unto him, Thou shalt love the Lord thy God with all thy heart, and with all thy soul, and with all thy mind. This is the first and great commandment" (Matt. 22: 36-38 KJV)."

Every time we Christians break a Commandment, our inner voice makes us feel guilty. For example, every time we dishonor God by disregarding the "great commandment" or each time we lie, cheat, steal, commit adultery, etc. against another person, we experience the same results: feelings of guilt and stress.

Repetition of disobedience creates chronic stress. Chronic stress is an unseen silent killer. (Stangor, C. and Walinga, J., 2014). It causes cardiovascular diseases (stroke, heart failure, and heart valve complication) and gastrointestinal disorders (constipation, nausea, abdominal pain, food poisoning, bloating, diarrhea, irritable bowel syndrome, and gastroesophageal reflux disease. Thus, disobedience to God's Commandments is a health risk for Christians. It is suicidal. Chronic stress is like drinking deadly poison and hoping that someone else will die from it. That's exactly what Jesus did. He died for our sins so that we do not have to die for them. He drank the "bitter cup" for us. All He asks of us, is to give Him our hearts/minds, so that He, through the Holy Spirit, can transform them into His likeness.

Let us learn to hate what God hates. Then we will recognize our own inability to make ourselves righteous, and fall before Jesus, pleading for a new heart of His making (see Ezekiel 36:26). When a student cheats on an examination and graduates, the guilt can stay with the student for the rest of the student's life. This guilt is a source of permanent stress if unconfessed. All the unconfessed sins of a Christian are a source of chronic stress, a silent killer. However, "If we confess our sins, He is faithful and just to forgive our sins and to cleanse us from all unrighteousness" (1 John 1:9 ESV). Spiritual Transformation occurs when we recognize that "all our righteousness is as filthy rags" (Isaiah 64:6 MEV) and we can only learn to hate what God hates by inviting Him to take control of our hearts and minds, remembering that He who promised to forgive our sins and cleanse us from all unrighteousness is FAITHFUL and JUST!

Figure 16 identifies seven things that the LORD hates: "These six things the Lord hates, yes, seven are an abomination to him: a proud look, a lying tongue, and hands that shed innocent blood, a heart that devises wicked imaginations, feet that are swift in running to mischief, a false witness who speaks lies, and he who sows discord among brethren" (Proverbs 6: 16-19 MEV). All these seven things that are hated by the LORD are evil, and yet, we are likely guilty of all or some of them, but Jesus has promised: "And I will give you a new heart—I will give you new and right desires—and put a new spirit within you. I will take out your stony hearts of sin and give you new hearts of love" (Ezekiel 36:26 TLB). "LORD Jesus, give me that new heart!"

Figure 16 illustrates the list of what the LORD Jesus Christ hates and which we can also hate when we listen to the Holy Spirit, our guide in the transformation process: "But when He, the Spirit of truth, comes, He will guide you into all the truth..." (John 16:13 NASB).

Figure 16. Seven things that are an abomination to the LORD. (Proverbs 6:16-19)

"LORD make me hate what You hate and love what You love."

Neuroscience of Joy and Happiness: Antidote to Chronic Stress

In Romans 15:13, Paul says: "I pray that God, the source of hope, will fill you completely with joy and peace because you trust in Him. Then you will overflow with confident hope through the power of the Holy Spirit" (NLT). This is an invitation to invoke the power of the Holy Spirit to give us the feelings of joy, peace, and hope.

Scientists report that when we engage in practices that increase feelings of happiness and joy, we simultaneously increase activity in our brain's left prefrontal cortex. As we continue to feel happy, we strengthen this activity and solidify brain pathways that make it easier to replicate feelings of happiness and joy. I propose that these feelings of happiness and joy are strongest and most long-lasting when guided by the Holy Spirit. Studies show that optimistic people have more activity in their left prefrontal cortex than pessimistic people. When Paul says, "but be ye transformed by the renewing of your mind," he means that as the Holy spirit renews our mind, we are transformed more and more into the likeness of Christ in thought, behavior, attitude, and actions, and our new neurons fire together and get wired together, giving us a mindset consistent with our desire to please God.

All the neural pathways formed and maintained by the Holy Spirit are stored in the brain and strengthen as we repeatedly reference them. As an example, consider the following: if devoted followers of Jesus Christ, listening to the Holy Spirit, die daily to self, then they are always in a mindset that says: God created all people in His image

- I love every image of God
- "Humble me, LORD, to be like You,"

The Holy Spirit will remold the brain, like the Potter and the clay, to be more like Christ and He will rewire its neurons into neural pathways that are designed to perceive and treat other people as they would treat the image of God, namely, love people and exhibit humility. Repetition of these thought processes will develop habits of love and humility. They will become engraved and imprinted into the brain This gives perfect peace. "Abundant peace belongs to those who love your instruction; nothing makes them stumble" (Psalm 119:165 CSB).

Chapter 16
Iniquity In Leadership

"Always treat others as you would like them to treat you; that sums up the teaching of the Torah (Law) and the Prophets" (Matthew 7:12 CJB).

Every human being is endowed with the capacity to feel what others are feeling. In other words, humans have empathy that can be measured in the brain. Seeing another individual in pain triggers an empathetic response in the observer (Caspar, et al 2020; Keysers & Gazzola, 2019; Decety, 2011; Singer & Lamm, 2009; Krishnan, et al, 2016). Studies show that naturally, we share what others feel because their pain is mapped onto our pain systems in the brain (Lamm & Majdandzic 2015). This empathic phenomenon is related to mirror neurons which fire when we execute and or observe an action (Gallese et al 1996; Keysers 2011). This is what causes normal people to avert harming others. When people submit to the orders of an authority figure to inflict pain on another, the neural response associated with the perception of pain felt by the victim is reduced for the perpetrator in comparison with being free to choose which action to perform. (Caspar, et al, 2020).

Why do people obey orders (from authoritarians) to do terrible things to others? Why do some people fail to resist these immoral orders? Is it possible that they deny personal responsibility when following the orders of a superior, and consequently, their inborn empathy is weakened when they follow such orders?

Since we are born guilty and shaped in iniquity, neural pathways associated with empathy for others are innately diminished. Obeying orders received from an authority figure reduces the vicarious brain activation of the empathic neural system when witnessing the pain that one has delivered to the victim (Caspar et al, 2020). Brain activity in empathy-related regions was attenuated (reduced) in the coerced condition. In other words, empathy activation is reduced in those who obey orders to inflict pain (emotional or physical) on others. (Caspar, et al, 2020). Of significance is the fact that reduced activation in the coerced condition, is associated with viewing shocks more frequently.

There is a reduced feeling of responsibility when people are coerced. (Caspar, et. al 2020). Even in the case of pain that is fully caused by the coerced individual's actions, brain activity is altered by a lack of responsibility (Caspar, et al, 2020). Obeying the orders of authority has a stronger influence on the empathic response toward others' pain than simply following the instructions of a computer (Caspar, et al, 2020).

For the Christian, the question is: "Who is your ultimate authority?" Whose orders do you obey? Are orders that cause mental, emotional, and physical pain to others consistent with the commands of Jesus, who declared: "But I say to you, love your enemies, bless those who curse you, do good to those who hate you, and pray for those who spitefully use you and persecute you" (Matthew 5:44 NKJV). Jesus further stated: "Be perfect, therefore, as your heavenly Father is perfect." (Matthew 5:48, NRSV). In this passage, being perfect is symbolized by how we treat other people (Matthew 5:43-47). Our major problem is that we often operate independently of God.

We pray about an issue but proceed to follow our own inclinations, instead of waiting for direction from the LORD. We end up feeding on human praise and pursuing popularity as a priority. Spiritual Transformation is obedience to the LORD Jesus. Such obedience enhances the activity of the empathy-related brain regions.

Obeying an order to hurt relaxes our aversion to harming others. Thus, obeying orders has a remarkable influence on how people perceive and process others' pain. This explains how people's willingness to inflict moral transgressions on others is enhanced in ordered situations (Caspar, et al 2020). I believe that if actions on others are ordered by Jesus, the recipient of the order is least likely to commit moral transgressions. The LORD said, "You shall not take revenge nor bear any grudge against your neighbor, but you shall love your neighbor as yourself. I am the LORD" (Lev. 19:18 NIV). The LORD gave this command. The Holy Spirit acts as the catalyst to strengthen empathetic neural pathways for obedience to this command. Neuroplasticity allows for the enhancement of the neural pathways for empathy.

Chapter 17
The Neuroscience Of Leviticus 19:17, 18

"You must not harbor hatred against your brother…"
"You shall not take vengeance or bear a grudge against any of your people, but you shall love your neighbor as yourself: I am the LORD" (Leviticus 19:17 (HCSB) and 18 NIV).

The LORD gave us these directives so that we can be in good health. How is this so? Scientists have discovered that hating people, keeping grudges, and harboring the desire for revenge (payback) produce hormones (chemical messengers) of stress. When hatred, grudges, and desire to revenge are prolonged, the stress becomes chronic and can lead to such chronic diseases as diabetes, hypertension, and congestive heart failure. Thus, hating people, keeping grudges, and wanting to pay back can literally make a person sick. In other words, thinking about our problems with other people produces stress hormones that make us sick. It is alarming that we can get sick just by dwelling on hateful thoughts. Thus, every thought of hatred, grudges, and revenge contributes to poor health. The LORD's commands in Leviticus 19:17, 18 are not just prohibitions, As our Creator, He knows that our thought processes can either make us healthy or sick.

Learning requires us to forge new connections (neural pathways) in the brain. There is a neural pathway for everything learned. Thus, when you learn something new, neurons fire together and form a new community of connections. Recalling serves as a reinforcement of these connections. Neural pathways for keeping grudges are reinforced every time the thought comes to mind. Neural connections for taking revenge get stronger each time the thought crosses your mind because neurons that fire together become stronger. Because the mind is the brain in action, thoughts of hatred, grudges, and revenge easily become dominant, controlling your actions every time the victim of your grudges and desire for revenge comes to mind. However, we can change our minds about keeping grudges, hating people, or wanting to revenge.

The LORD designed a unique microscopic shuttle to occur when we decide to change our minds (thoughts). It is the first law of neuroplasticity that states: "Neurons that fire together wire together." The more we focus on harboring hatred, the stronger the neural circuit of hatred. The second law of neuroplasticity states: "Neurons that fire apart wire apart." For example, if you deliberately fire into your brain thoughts of compassion and love for people, your grudge/hate neurons disconnect (fire apart), and are supplanted by the thoughts of love and compassion. This is difficult for unconverted individuals to do, because they cherish their hateful thoughts. However, the Holy Spirit is very capable of re-wiring the minds of those who submit to Him. He strengthens the new neural connections and supplants the old hateful ones. He biologically and neurologically trims away the old mind, replacing it with a Christlike mind and spirit, where love prevails. Then the commands of God will be written on the heart (mind) as the LORD promised Israel in Jeremiah 31:33: "I will put My law in their minds, and **write** it on their **heart**s…" (NKJV). This may not mean that the person totally forgets. It simply means that the spiritually transformed person no longer cherishes grudges and hatred. Paul tells us, "...take captive every thought to make it obedient to Christ" (2 Cor. 10:5 NIV). When this is done repeatedly, the Holy Spirit replaces the old brain circuits that controlled you in the past.

Neuroinfluence™ and Spiritual Transformation

Philippe Douyon, CEO of Inle Brainfit Institute, coined the term Neuroinfluence™ (https://www.inlebrainfitinstitute.com/). According to Dr. Douyon, neuroinfluence is everything that impacts our brains, structure, development, and function. The impact of Neuroinfluence™ includes all the things that (a) promote the formation and death of neurons, and (b) cause new neural pathways to develop and old ones to die. It refers to those things that encourage synapses (interneural connections) to form, neurotransmitters (chemical messengers) to be released, neurons to fire and wire together, and that cause connections (neural pathways) to strengthen, weaken or disappear. Dr. Douyon describes neuroinfluence™ as everything that our brains have ever and will ever encounter and the effect of those encounters on neuroplasticity (Douyon, 2019). The human brain was created by God to respond to all stimuli of its neuroinfluence™ by either giving birth to new neurons, creating new connections, strengthening/weakening the networks which already exist, or eliminating (pruning) neurons programmed for elimination.

As human beings, we have a major impact on our neuroinfluence™ through self-talk and mindfulness and mindsight, as described earlier. Spiritual transformation occurs when a person submits fully to the Holy Spirit who creates a neuroinfluencial environment that is pleasing to our LORD Jesus Christ. "A joyful heart (Christ-centered mindset) is good medicine, but a crushed spirit dries up the bones (Proverbs 17:22, KJV). A spiritually transformed heart (person) hates what God hates (Figure 17).

Holy Spirit Directed Spiritual Transformation and Neuroplasticity

As a Christian, what one thing would you want to change in your life to be more like Christ? The LORD God created our brains to be neuroplastic, allowing every brain to be malleable and subject to physical, structural, and functional changes. This requires us to intentionally redirect our focus and attention. The Apostle Paul admonishes us: "[6] Do not be anxious about anything, but in every situation, by prayer and petition, with thanksgiving, present your requests to God. [7] And the peace of God, which transcends all understanding, will guard your hearts and your minds in Christ Jesus. [8] Finally, brothers and sisters, whatever is true, whatever is noble, whatever is right, whatever is pure, whatever is lovely, whatever is admirable—if anything is excellent or praiseworthy—think about such things" (Philippians 4:6-8 NIV). Many people, even Christians, have a hard time meeting this standard.

Before His arrest and crucifixion, Jesus gave precious promises to His disciples that apply to us also. He said: "And I will pray the Father, and He will give you another Helper, that He may **abide with you forever**— [17] the Spirit of truth, whom the world cannot receive, because it neither sees Him nor knows Him; but you know Him, for He dwells with you and **will be in you**" (John 14:16, 17 NKJV). In verse 26 of the same chapter, Jesus adds: "But the Helper, the Holy Spirit, whom the Father will send in My name, He **will teach you all things**, and bring to your remembrance all things that I said to you." In chapters 15 and 16 of John, Jesus provides further information about the work of the Holy Spirit. "But when the Helper comes, whom I shall send to you from the Father, the Spirit of truth who proceeds from the Father, **He will testify of Me**" (John 15:26 NKJV); "…for if I do not go away, the Helper will not come to you; but if I depart, I will send Him to you. And when He has come, He **will convict the world of sin, and of righteousness, and of judgment**" (John 16:7, 8 NKJV); "…when He, the Spirit of truth, has come, He **will guide you into all truth**… He will glorify Me…" (John 16:13, 14 NKJV).

Under the influence of the Holy Spirit, our brains can be rewired to overcome limiting beliefs, negative thinking patterns, disruptive emotions, self-sabotaging behaviors, as well as prejudices. The Holy Spirit helps us to focus our attention on the Biblical LORD Jesus, His teachings, His sacrifice, and His redeeming love and grace. Jesus talks to us through the Bible and through the inner voice (the Holy Spirit that He promised to send to us) warns us when we do wrong. This intentional focus on Jesus, buttressed by Christ-centred self-talk, allows for the rewiring of our neural pathways. When this pattern is repeated time and again, our brain creates permanent neural networks that displace the old networks. We become new creatures in Christ. Spiritual transformation occurs when we submit to Christ-centered, Holy Spirit directed neuroplasticity reformation, which is an intentional process resulting in a physical change in our brain

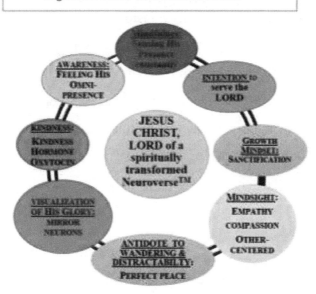

Figure 17. Christ-Centered Neuroverse™

structures and functions as the result of focused attention on the LORD Jesus Christ as the pivotal center of neuroinfluence™. Only then can we hope to acquire the "mind" that is in Christ Jesus. Figure 17 lists eight principles embodied in the Neuroverse™. I propose that spiritual transformation occurs when the LORD Jesus is at the Center of the Neuroverse™. A spiritually transformed Christian's life is dominated by each one of the components of the Neuroverse™ in Figure 17.

"There is nothing more dangerous or tragic than the failure of professing Christians to keep Christ central in all things. The Christian life is all about the supremacy of Christ." Brian G. Hedges in *Centrality of Christ in Christianity* (https://www.christianstudylibrary.org/article/centrality-christ-christianity).

I propose that there is a guiding principle for keeping Christ central in the life of a spiritually transformed Christian. That principle is: always operate on the basis and understanding that any situation is predicated and grounded on 'Not I, but Christ.' This spirit of selflessness should be displayed in our marriages, occupations, money-management, recreation, church activities, classroom instruction, and relationships with young and old people, and all other relationships. The <u>'Not I, but Christ'</u> principle should dominate the *Neuroverse*™ as evidence of a spiritually transformed Christian.

The pre-requisite for Spiritual Transformation is complete surrender to the LORD Jesus Christ who created the neuroplastic brain. (See Genesis 1:26, 27 and John 1:1-3, 14). Therefore, He is the engineer that must be consulted since spiritual transformation is rewiring our brain to be Christ-like in all that we do and think. Studies have shown that it takes an average of 66 days to permanently acquire a new habit. Jesus told Peter, "When thou art converted, strengthen thy brethren" (Luke 22:32, KJV). It is necessary to take micro (small) steps to break undesirable habits and replace them with good ones. This is critical for the successful displacement of old habits with new ones. Each small step makes the brain produce new neurons that subsequently form neural pathways for the desired behavior, such as spiritual transformation.

The Bible does not leave us without instruction in righteousness. In Proverbs 3:5-6, we find, "Trust in the Lord with all your heart, and do not lean on your own understanding. In all your ways acknowledge him, and he will make straight your paths (ESV). Isaiah reminds us: "For thus says the One who is high and lifted up, who inhabits eternity, whose name is Holy: 'I dwell in the high and holy place, and also with him who is of a contrite and lowly spirit, to revive the spirit of the lowly, and to revive the heart of the contrite'" (Isa 57:15, ESV).

Jeremiah was a young man when called by the LORD God to be a prophet. The LORD identified Himself to Jeremiah as both Creator and LORD. "The word of the **LORD** came to me, saying, 'Before **I formed you** in the womb I knew you, before you were born I set you apart; I appointed you as a prophet to the nations.'" (Jeremiah 1:4, 5, NIV). The LORD is both the Creator and the engineer of our brains. He created within us a neuroplastic brain with the potential to be dedicated to serving Him. The discovery of the brain's ability to undergo neuroplasticity throughout the lifespan has opened our eyes to the possibilities associated with ongoing transformation. It means that when we desire Spiritual Transformation, we request Jesus to facilitate our taking the micro-steps necessary for the rewiring of our neural pathways and we allow Him to take Lordship in our lives. Jesus says "…apart from me you can do nothing" (John 15:5 NIV). The Creator/Engineer and His divine associate, the Holy Spirit, can rewire our neural pathways which facilitates His transformation of our hearts and minds to love Him and our fellowmen.

When the LORD Jesus says, "Abide in Me and I in you" (John 15:4 NIV), He is inviting us to focus intentionally and deliberately on Him and listen to the Holy Spirit whom He sent to "guide us into all truth". In Colossians 1:25-27, Paul provides a heart-touching revelation of what Jesus desires for each of us: "God has sent me to help his Church and to tell his secret plan to you Gentiles. He has kept this secret for centuries and generations past, but now at last it has pleased him to tell it to those who love him and live for him, and the riches and glory of his plan are for you Gentiles, too. And this is the secret: *Christ in your hearts is your only hope of glory"* (TLB). Jesus spoke to this point when asked about the greatest commandment. "Jesus answered him, 'You shall love the Lord your God with all your heart, with all your soul, and with all your mind'. This is the first and great commandment. And there is a second like it: 'You shall love your neighbour as yourself'. The whole of the Law and the Prophets depends on these two commandments" (Matthew 22:38 Phillips). There is a close connection between the heart and the mind.

The spiritual transformation process involves the rewiring of our brains which includes the production of neural pathways for dependence on Jesus. When we focus on Him and pay attention to His word, the Holy Spirit is able to engrave permanent God-pleasing neural pathways in the brain. The outcome is a Christian whose focus is on Jesus, who listens attentively to the Holy Spirit, and who desires to please the LORD, no matter what. This is how neuroscience connects with spiritual transformation. This Spirit-directed and God-focused neuroplasticity can occur at any time in the life of any individual who makes a commitment to Jesus. Figure 18 is a road map to spiritual transformation.

Figure 18. Summary of a Christ-Centered Neuroverse™ : Road Map for Spiritual Transformation

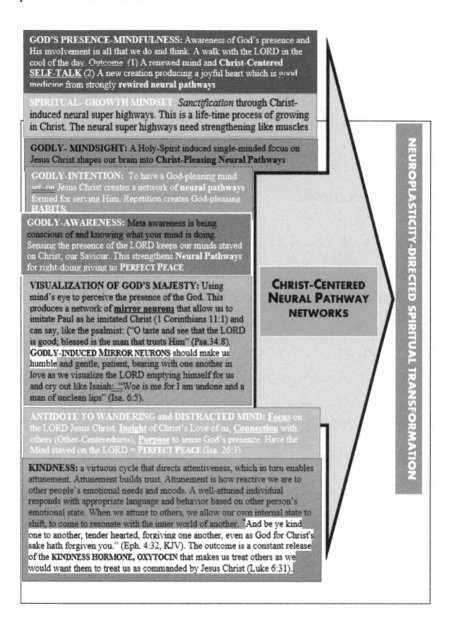

References

Ackerman, C.E. (2021) *What is Gratitude and Why is it so Important?-*

Agarwal, P. (2020) *Exposing unconscious bias - https://bit.ly/2Vqu6TP.*

Amen, D. (1996) Change Your Brain, Change Your Life - https://amzn.to/3jJnuKl

Ashton, M. C., & Lee, K. (2007). Empirical, theoretical, and practical advantages of the HEXACO model of personality structure. *Personality and Social Psychology Review, 11*(2), 150–166. https://doi.org/10.1177/1088868306294907. (Published in final edited form as: Neuroimage. 2012 March ; 60(1): 683–692. doi:10.1016/j.neuroimage.2011.12.069.)

Begley, S. (2007) Train Your Mind, Change Your Brain - https://amzn.to/3hlvv6v

Bosman, M., (2021) *(https://www.stratleader.net/sli-blog/self-directed-neuroplasticity) August 18, 2021. "Self-Directed Neuroplasticity: Change Your Life by Changing Your Focus" by Monie Bosman (https://www.stratleader.net/sli-blog/self-directed-neuroplasticity) August 18, 2021*

Bosman, M. (2012). *You Might Not Like it, But Bad is Stronger than Good -https://bit.ly/2W3bJEQ).*

Burke & Barnes (2006) *Neural plasticity in the ageing brain* - *https://bit.ly/3xIUgyz*

Caspar, et al, (https//:doj.org/10.1016/j.neuroimage.2020.117251).

Cherry, K. (2020) *How We Use Selective Attention to Filter Information and Focus* - *https://bit.ly/3xX8ufg*

Colin Brown, The New International Dictionary of New Testament Theology (Grand Rapids: Zondervan, 1967), 2:259.

Blackwell, L.S.; Trzesniewski, K.H.; Dweck, C.S. Implicit theories of intelligence predict achievement across an adolescent transition: A longitudinal study and an intervention. Child Dev. 2007, 78, 246–263.

Cherry, K. (2020) *How We Use Selective Attention to Filter Information and Focus* - *https://bit.ly/3xX8ufg*

Moore, C. (2021) *What Is Mindfulness?* - *https://bit.ly/3jXkKHL*

Ding, K., Liu, J. (2022). Comparing gratitude and pride: evidence from brain and behavior. *Cogn Affect Behav Neurosci* (2022). https://doi.org/10.3758/s13415-022-01006-y)

Douyon, 2019, page 81. [Neuroplasticity: Your Brain's Superpower, published by Izzard Publishing (ISBN: 978-1-64228-100-2)].

Doidge, N. (2007) *The Brain That Changes Itself* - https://amzn.to/3jM3Xcb

Dweck, C.S. (2007). Mindset: The New Psychology of Success. Ballantine Book Publishers

ISBN 0345472322.

Dweck, C. S. (2008). Mindset: The New Psychology of Success. Random House; New York

E. G. White: Be ye perfect. *Letters and Manuscripts.* Vol 12 paragraph 14, 1897)

Gainsburg & Kross (2020). Distanced self-talk changes how people conceptualize the self - https://bit.ly/3jX3WAG).

Gardner, Benjamin. (2012). Making health habitual: the psychology of habit formation and General Practice. Br. J. Gen. Pract. 62(603):664-666.

Glicksman, E. (2019) (https://www.brainfacts.org/Thinking-Sensing-and-Behaving/Emotions-Stress-and-Anxiety/2019/Your-Brain-on-Guilt-and-Shame-091219

The Neuroscience of Shame: https://cptsdfoundation.org/2019/04/11/the-neuroscience-of-shame/

Hanson, R. (2015) Just One Thing: Pay Attention! https://bit.ly/3k4r0PD

Hanson, R. (2013) How to Grow the Good in Your Brain - https://bit.ly/3hiYSXd

Hollins, P. Build Better Brain: Using Neuroplasticity to train your brain for motivation, discipline, courage and mentala sharpness. www.petehollins.com

Keysers, Christian (2010). *"Mirror Neurons" (PDF)*. Current Biology. *19* (21): R971–973 *doi:10.1016/j.cub.2009.08.026*. PMID *19922849*.

Mirror Neurons Differentially Encode the Peripersonal and Extrapersonal Space of Monkeys

Kilingsworth, A & Gilbert DT (2010). A wandering mind is an unhappy mind. *Science* Vol. 330, p. 932. *https://bit.ly/3g9Crmm)*

Kim et al (2021). *The effects of positive or negative self-talk on the alteration of brain functional connectivity by performing cognitive tasks* - *https://bit.ly/3g8Pvbv*

Lewis C.S. Mere Christianity New York: Simon & Schuster Touchstone Edition 1996, pp. 109-112).

Lutz, A, Brefczynski-Lewis J, Johnstone T, Davidson RJ (2008) Regulation of the Neural Circuitry of Emotion by Compassion Meditation: Effects of Meditative Expertise. PLoS ONE 3(3): e1897. https://doi.org/10.1371/journal.pone.0001897)

Moore, C. *(2021) What Is Mindfulness?* - *https://bit.ly/3jXkKHL*

Ng, Betsy (2018) The Neuroscience of Growth Mindset and Intrinsic Motivation. *Brain Sciences* **8**:20 (doi:10.3390/brainsci8020020).

Nigel Mathers: Compassion and the science of kindness: Harvard Davis Lecture 2015 by *British Journal of General Practice* 2016; 66 (648): e525-e527. DOI: https://doi.org/10.3399/bjgp16X686041

Phillippa, L., Cornelia, vanJaarsveld, Potts Henry and Warde, Jane (2009). European Journal of Social Psychology. **40 (6)**: 998-1009.

Rajendra A. Morey, Gregory McCarthy, Elizabeth S. Selgrade, Srishti Seth, Jessica D. Nasser, and Kevin S. LaBar *NeuroImage* 60 (1): March 2012, pp 683-692.

Rizzolatti, Giacomo; Craighero, Laila (2004). *"The mirror-neuron system" (PDF)*. *Annual Review of Neuroscience*. 27 (1): 169–192. doi:*10.1146/annurev.neuro.27.070203.144230*. PMID

Schmelzer, (2015) http://gretchenschmelzer.com/blog-1/2015/1/11/understanding-learning-and-memory-the-neuroscience-of-repetition

Schwartz & Begley (2003) The Mind and the Brain: Neuroplasticity and the Power of Mental Force – https://amzn.to/36NHsfh)

Stangor, C. and Walinga, J. (2014). *Introduction to Psychology – 1st Canadian Edition*. Victoria, B.C.: Campus. Retrieved from https://opentextbc.ca/introductiontopsychology

Tarrants, Thomas A (2011) "Pride and Humility," Knowing and Doing, C. S. Lewis Institute, Winter 2011, http://www.cslewisinstitute.org/Pride_and_Humility_SinglePage, accessed

Tang et al (2015) *The neuroscience of mindfulness meditation -* *https://bit.ly/3mdtnk7*

Tracy, J. L., & Robins, R. W. (2004a). Putting the self into self-conscious
emotions: A theoretical model. *Psychological Inquiry,15(2)*, 103–125.

Tracy, J. L., & Robins, R. W. (2004b). Show your pride: Evidence for a discrete emotion expression. *Psychological Science,* 15(3), 194–197.

Tracy, J. L., & Robins, R. W. (2007a). Emerging insights into the nature and function of pride. *Current Directions in Psychological Science,* 16(3), 147–150.

Tracy, J. L., & Robins, R. W. (2007b). The psychological structure of pride: A tale of two facets. *Journal of Personality and Social Psychology,* 92(3), 506.

Tracy, J. L., & Robins, R. W. (2008). The automaticity of emotion recognition. *Emotion,* 8(1), 81.

Tracy, J. L., Shariff, A. F., & Cheng, J. T. (2010). A naturalist's view of pride. *Emotion Review,* 2(2), 163–177.

Widdett, R. (2014*) Neuroplasticity and Mindfulness Meditation - https://bit.ly/3kA8Y7Z*

Wheeler et al (2017) *The Neuroscience of Mindfulness: How Mindfulness Alters the Brain and Facilitates Emotion Regulation - https://bit.ly/37OihK*

Made in the USA
Monee, IL
27 January 2023